Scientific explanation

Scientific explanation

PAPERS BASED ON HERBERT SPENCER LECTURES GIVEN IN THE UNIVERSITY OF OXFORD

EDITED BY
A. F. HEATH
Fellow of Jesus College, Oxford

CLARENDON PRESS · OXFORD
1981

Oxford University Press, Walton street, Oxford OX2 6DP
London Glasgow New York Toronto
Delhi Bombay Calcutta Madras Karachi
Kuala Lumpur Singapore Hong Kong Tokyo
Nairobi Dar es Salaam Cape Town
Melbourne Auckland
and associate companies in
Beirut Berlin Ibadan Mexico City

Published in the United States by
Oxford University Press, New York

British Library Cataloguing in Publication Data
Scientific, explanation.–(Herbert Spencer lectures)
1. Science–Philosophy–Addresses, essays, lectures
I. Title II. Heath, A. F. III. Series
501 Q175.3
ISBN 0-19-858214-5

Set by Hope Services, Abingdon
and printed in Great Britain by
Billing & Son Ltd., Guildford

Preface

The Herbert Spencer lectures began in 1905 and over the following seventy-five years have covered a rich *melange* of topics which have borne a more or less close relationship to Herbert Spencer's own intellectual interests. Francis Galton lectured on 'probability, the foundation of eugenics'; Lucien Levy Bruhl on 'primitive mentality'; Sir William Beveridge on 'the economic aspects of planning under socialism'; F. C. Bartlett on 'thinking'; J. B. S. Haldane on 'time in biology'; Morris Ginsberg on 'social change'; and P. B. Medawär on 'evolution and evolutionism'.

One recurrent theme throughout the series has been the philosophy of science. Indeed, it is perhaps the one with the most distinguished tradition of lecturers including as it does Bertrand Russell, Albert Einstein, and Karl Popper. Einstein himself said at the beginning of his Herbert Spencer lecture that 'if you want to find out anything from the theoretical physicists about the methods they use, I advise you to stick closely to one principle: don't listen to their words, fix your attention on their deeds'. The present volume of 1979 lectures enables the reader to judge the wisdom, or otherwise, of Einstein's advice. It brings together accounts by distinguished scientists, historians, and philosophers of science about the nature of scientific explanation.

We have three practising scientists—Abdus Salam, himself a Nobel prizewinner for theoretical physics, the psychologist Donald Broadbent, and Henry Harris, the Regius Professor of Medicine at Oxford. Harris concludes with a thumb-nail sketch of the practising scientist: 'He is a thoroughgoing empiricist who never troubles his head about the logic of what he is doing, but has no doubt that his activities yield information about the real world.' Absence of doubt, and suspension of disbelief, figure prominently in the lecture given by Gerald Holton, a historian of science (as well as a distinguished physicist in his own right). He gives us a fascinating account of the deeds, and lack of doubt, of the early twentieth-century physicist, R. A. Millikan, based on the latter's laboratory notebooks.

Our two philosophers are Daniel Dennett and Hilary Putnam. The certainty of the practising scientists contrasts with the uncertainty of the philosopher about the account which has to be given of their activities. Thus Putnam argues that both of the two most influential philosophies of science of the twentieth century—logical positivism and the Kuhn–Feyerabend thesis of incommensurability—are self-refuting. (As he points out, 'To tell us that Galileo had "incommensurable" notions *and then go on* to describe them at length is totally incoherent.') In their place he offers us a somewhat, but not wholly, precarious view of science and culture. It is the image of a fleet of boats. 'The people in each boat are trying to reconstruct their own boat without modifying it so much at any one time that the boat sinks . . . In addition, people are passing supplies and tools from one boat to another and shouting advice and encouragement (or discouragement) to each other. Finally, people sometimes decide they do not like the boat they are in and move to a different boat altogether. (And sometimes a boat sinks or is abandoned.)' The present volume is a part of that process of shouting advice and encouragement (or discouragement) from one boat to another.

Oxford A. F. H.
April 1981

Contents

LIST OF CONTRIBUTORS ix

1. THEMATIC PRESUPPOSITIONS AND THE DIRECTION OF SCIENTIFIC ADVANCE
Gerald Holton 1

2. THE NATURE OF THE 'ULTIMATE' EXPLANATION IN PHYSICS
Abdus Salam 28

3. RATIONALITY IN SCIENCE
Henry Harris 36

4. TRUE BELIEVERS: THE INTENTIONAL STRATEGY AND WHY IT WORKS
Daniel C. Dennett 53

5. NON-CORPOREAL EXPLANATION IN PSYCHOLOGY
Donald E. Broadbent 76

6. PHILOSOPHERS AND HUMAN UNDERSTANDING
H. Putnam 99

INDEX 121

List of contributors

Donald Broadbent is a member of the Medical Research Council External Staff and works in the Department of Experimental Psychology at Oxford University. He is a former President of the British Psychological Society and of the Experimental Psychology Society, a Fellow of the Royal Society and a Foreign Associate of the US Academy of Sciences. In 1975 he received the Distinguished Scientific Contribution Award of the American Psychological Association. He now has well over a hundred published scientific papers, but his main books are *Perception and communication; Behaviour; Decision and stress* and *In defence of Empirical psychology*. He is a Fellow of Wolfson College.

Daniel C. Dennett is Professor and Chairman of the Philosophy Department at Tufts University. He is currently the President of the Society for Philosophy and Psychology and has been a Visiting Professor at Harvard University, a Fullbright Fellow at Bristol University and a Fellow at the Center for Advanced Study in The Behavioural Sciences at Stanford, where he worked with a special research group on artificial intelligence and philosophy. He is the author of *Content and consciousness* and of *Brainstorms*.

Henry Harris is the Regius Professor of Medicine at Oxford and Head of the Sir William Dunn School of Pathology. He has made many contributions to cell biology, but is perhaps best known for his discovery of the short-lived RNA in the cell nucleus and for his use of the technique of cell fusion to provide new approaches to basic problems in the genetics and physiology of somatic cells. He is a Fellow of the Royal Society, a Foreign Honorary Member of the American Academy of Arts and Sciences, a Foreign Member of the Max Planck Society and an

Honorary Fellow of the Cambridge Philosophical Society. He is the author of *Nucleus and cytoplasm* and *Cell fusion*.

Gerald Holton is Mallinckrodt Professor of Physics and Professor of the History of Science at Harvard University. Among other honours he has been selected as the Thomas Jefferson Lecturer for 1981 of the National Endowment for the Humanities. His research interests are physics of high pressure phenomena and the history of physical science. His publications include *Thematic Origins of Scientific Thought: Kepler to Einstein* and *The Scientific Imagination: Case Studies*.

Hilary Putnam is Walter Beverly Pearson Professor of Modern Mathematics and Mathematical Logic and Chairman of the Department of Philosophy, at Harvard University. He is a Fellow of the American Academy of Arts and Sciences, Corresponding Fellow of the British Academy and has been President of the Philosophy of Science Association and of the Association for Symbolic Logic. He is author of *Philosophy of logic*; *Mathematics, matter and method*; *Language, mind, and reality*; and most recently of *Meaning and the moral sciences*.

Abdus Salam is Professor of Theoretical Physics at Imperial College, London and Director of the International Centre for Theoretical Physics, Trieste. He is a Fellow of the Royal Society and received the Nobel Prize for Physics in 1979. He has published over 200 scientific papers on physics of elementary particles as well as papers on scientific and educational policy for Pakistan and developing countries.

1 *Thematic presuppositions and the direction of scientific advance*

GERALD HOLTON

Jefferson Laboratory, Harvard University

'I wish to preface what I have to say by expressing to you the great gratitude which I feel to the University of Oxford for having given me the honour and privilege of delivering the Herbert Spencer Lecture.'

With these words, surely echoed by every speaker in this series, Albert Einstein opened his lecture on 10 June 1933. By that time he was a man without a country, passing through this haven as a refugee from Fascism, as so many others, illustrious or unknown, were to do after him. Like them, he retained a warm and thankful memory of the hospitality here.

Philipp Frank, his biographer and colleague, called Einstein's lecture the 'finest formulation of his views on the nature of a physical theory'.[1] The published version[2] has been rarely analysed or even adequately understood. Now that we have access to so many more of Einstein's published and unpublished documents, the essay turns out to be a very appropriate entry for a study of scientific explanation, both of Einstein's own contribution to the subject and of more recent approaches.

The 'eternal antithesis'

Einstein's choice of 'the method of theoretical physics' as his topic was by no means casual. In fact, for much of his life he seems to have been almost obsessed by the need to explain what he called his epistemological credo. From about 1911 to the end, he wrote on it again and again, almost as frequently as on physics itself. On occasions great and small, he reverted to his self-appointed task in his remarkably consistent way—with the single-minded patience of a hedgehog, and the glorious stubbornness that characterized him from his boyhood on, when his family watched him at one of his favourite activities, making with infinite concentration fantastic houses of cards that had as many as ten levels.

His home-made philosophical system of the practising scientist, of which he wrote so often, seemed to his philosophical commentators something of a house of cards too, a patchwork of pages from Hume, Kant, Ernst Mach, Henri Poincaré, and many others. Indeed, Einstein himself cheerfully acknowledged once that he might appear 'as a type of unscrupulous opportunist', appearing by turns as a realist, idealist, positivist, or even Platonist or Pythagorian. Yet the method he preached and practised turned out to be remarkably robust. Many of today's physicists, without knowing its origin, have adopted a style of attempting fundamental and daring advances that owes a great deal to Einstein's credo, even as Einstein's dream of finding a unification of the forces of nature has, in its modern form, turned out to be the stuff of which Nobel prizes are made.

In his own day, however, Einstein had good reason to suspect that few physicists and philosophers understood what he was saying about scientific methodology, or even could describe clearly what they themselves were doing. And so, rather like Galileo, he took his epistemological message to the wider public. He opened the formal part of his Herbert Spencer lecture with the famous sentence: 'If you want to find out anything from the theoretical physicists about the methods they use, I advise you to stick closely to one principle: don't listen to their words, fix your attention on their deeds.'

Here he objects to scientists who speak about the products of their imaginations as if these were 'necessary and natural'—not 'creations of thought' but 'given realities'. To expose their mistake, he invites us to pay 'special attention to the relation between the content of a theory' on the one hand, and 'the totality of empirical facts' on the other. These constitute the two 'components of our knowledge', the 'rational' and the 'empirical'; these two components are 'inseparable'; but they stand also, Einstein warns, in 'eternal antithesis'.

To support this conception, Einstein now gives a very brief sketch of a dichotomy built into Western science. The Greek philosopher-scientists provided the necessary confidence for the achievements of the human intellect by introducing into Western thought the 'miracle of the logical system', which, as in Euclid's geometry, 'proceeds from step to step with such precision that every single one of its propositions was absolutely indubitable'. But 'propositions arrived at by purely logical means are completely empty as regards reality'; 'through purely logical thinking we can attain no knowledge whatsoever of the empirical world'. Einstein tells us that it required the seventeenth-century scientists to show that scientific knowledge 'starts from experience and ends with it'.

It seems therefore that we are left with a thoroughly dualistic method for doing science: on the one hand, Einstein says, 'the structure of the system is the work of reason'; on the other hand, 'the empirical contents and their mutual relations must find their representation in the conclusions of the theory'. Indeed, virtually all of Einstein's commentators have followed him in stressing this dualism—and leaving it at that. For example, F. S. C. Northrop summarized the main content of Einstein's Oxford lecture in these words: An 'analysis of Einstein's conception of science shows that scientific concepts have two sources for their meanings: The one source is empirical. It gives concepts which are particulars, nominalistic in character. The other source is formal, mathematical and theoretical. It gives concepts which are universals, since they derive their meaning by postulation from postulates which are universal propositions.'[3]

This is a view of science (even of Einstein's science) of which there are many versions and variants. I would call it a two-dimensional view. It can be defended, up to a point. All philosophies of science agree on the meaningfulness of two types of statements, namely propositions concerning empirical matters that ultimately boil down to meter readings and other public phenomena, and propositions concerning logic and mathematics that ultimately boil down to tautologies. The first of these, the propositions concerning empirical matters of fact, can in principle be rendered in protocol sentences in ordinary language that command the general assent of a scientific community; I like to call these the *phenomenic propositions*. The second type of propositions, meaningful in so far as they are consistent within the system of accepted axioms, can be called *analytic propositions*. As a mnemonic device, and also to do justice to Einstein's warning about the 'eternally antithetical' nature of these propositions, one may imagine them as lying on a set of orthogonal axes, representing the two dimensions of a plane within which scientific discourse usually takes place.

Now it is the claim of most modern philosophies of science which trace their roots to empiricism or positivism, that any scientific statement has 'meaning' only in so far as it can be shown to have phenomenic and/or analytic components in this plane. And indeed, in the past, this Procrustean criterion has amputated from science its innate properties, occult principles, and all kinds of tantalizing questions for which the consensual mechanism could not provide answers. A good argument can be made that the silent but general agreement to keep the discourse consciously in the phenomenic-analytic plane where statements can be shared and publicly verified or falsified is a main reason why science has been able to grow so rapidly in modern times. The same approach also characterizes the way science is taught in most classrooms, and is 'rationalized' in most of the current epistemological discussions.

Problems for the two-dimensional view

Nevertheless, this two-dimensional view has its costs. It overlooks or denies the existence of active mechanisms at work in the day-to-day experience of those who are actually engaged in the pursuit of science; and it is of little help in handling questions every historian of science has to face consciously, even if the working scientist, happily, does not. To illustrate, let me mention two such puzzles. Both have to do with the direction of scientific advance, and both will seem more amenable to solution once the dualistic view is modified.

1. If sound discourse is directed entirely by the dictates of logic and of empirical findings, why is science not one great totalitarian engine, taking everyone relentlessly to the same inevitable goal? The laws of reason, the phenomena of physics, and the human skills to deal with both are presumably distributed equally over much of the globe; and yet the story of, say, the reception of Einstein's theories is strikingly different in Germany and England, in France and the United States. On the level of *personal* choice of a research topic, why were some of Einstein's contemporaries so fatally attracted to ether-drift experiments, whereas he himself, as he put it to his friend de Haas, thought it as silly and doomed to failure as trying to study dreams in order to prove the existence of ghosts? As to skills for navigating in the two-dimensional plane, Einstein and Bohr were rather well matched, as were Schrödinger and Heisenberg. And yet there were fundamental antagonisms in terms of programmes, tastes, and beliefs, with occasional passionate outbursts between scientific opponents.

Or, again, how to understand the great variety of different personal styles? The physicist Edwin C. Kemble described his typical mode of work, with some regret, as the building of a heavy cantilevered bridge, each piece painstakingly anchored on a well-secured base. Robert Oppenheimer, on the other hand, one might think of as a spider building a web; individual extensions were achieved by daring leaps, and the resulting structures were intricate and shimmering with beauty, but perhaps a bit fragile. Enrico Fermi, whom many regard as the inventor of teamwork in modern physics, ran his laboratory like a father who had assembled around himself a group of very bright offspring.

And then there is the scientist who moves through his problem-area alone, as the fur trapper did through Indian territory. Bernard DeVoto described it in his book *Across the wide Missouri*. The trapper 'not only worked in the wilderness. He also lived there. And he did so from sun to sun by the exercise of total skill'. Learning how to read formal signs was of course essential to him, but more important was 'the interpretation of observed circumstances too minute to be called signs. A branch floats down a stream–is this natural, or the work of animals, or of Indians or trappers? Another branch or a bush or even a pebble is out of place–why? . . . Buffalo are moving down wind, an elk is in an unlikely place or posture, too many magpies are hollering, a wolf's howl is off key–what does it mean?'

What indeed does all this variety of scientific styles mean? If science *were* two-dimensional, the work in a given field would be governed by a rigid, uniform paradigm. But the easily documented existence of pluralism at all times points to the fatal flaw in the two-dimensional model.

2. A second question that escapes the simple model, and to which I have devoted a number of case studies in recent years, is this: why are many scientists, particularly in the nascent phase of their work, willing to hold firmly, and sometimes at great risk, to a form of 'suspension of disbelief' about the possibility of falsification? Moreover, why do they do so sometimes without having any empirical evidence on their side, or even in the face of disconfirming evidence?

Among countless examples of this sort, Max Planck, responsible for the idea of the quantum but one of the most outspoken opponents of its corpuscular implications, cried out as late as May 1927 'Must we really ascribe to the light quanta a physical reality?–and this four years after the publication and verification of Arthur H. Compton's findings. On the other hand, when it came to explaining the electron in terms of what Planck called 'vibrations of a standing wave in a continuous medium', along the lines proposed by de Broglie and Schrödinger, Planck gladly accepted the idea and added that these principles have already [been] established on a solid foundation'–and all that before Planck had heard of any experimental evidence along the lines provided by Davisson and Germer.

'I do not doubt at all . . .'

Einstein was even more daring. As I have documented elsewhere, straight after the publication of his 1905 relativity paper there appeared what purported to be an unambiguous experimental disproof of it by the most eminent experimentalist in the field, Walter Kaufmann. If Einstein had been a naïve believer in such notions as falsification criteria or regressive research programmes, he would have had to accept this widely noted disproof from that undoubted source, and turned to other things. For the published data showed that the electrons' motion fitted ether-based theories far better than Einstein's. Yet Einstein paid no attention whatever, and continued to publish as if nothing had happened. When the young man was finally persuaded to respond to the challenge, he dismissed the supposed disproof with a characteristic declaration: The ether-based theories 'have a rather small probability, because their fundamental assumptions concerning the mass of moving electrons are not explainable in terms of theoretical systems which embrace a greater complex of phenomena.' (It took ten years for it to be fully realized that, for once, the prominent experimenter had been working with quite inadequate equipment. By that time, the matter had been settled on other grounds, as it is so often.)

Later, when the gravitational red shift, predicted by general relativity theory for the spectral lines from stars with large masses, turned out to be very difficult to test, and the experimental results were neither systematic nor of the predicted amount, Einstein again simply waited it out. To Max Born he wrote later that, even in the absence of all three of the originally expected observable consequences of general relativity, his central gravitation equations 'would still be convincing', and that in any case he deplored that 'human beings are normally deaf to the strongest [favourable] arguments, while they are always inclined to overestimate measuring accuracies'.

To be sure, if one looks hard, one can find in Einstein's voluminous writings a small number of statements of the opposite kind. An example of this sort, written shortly after the triumphant announcement of Eddington's results late in 1919, is one sentence in the 1920 edition of Einstein's popular exposition, *Relativity, the special and general theory*: 'If the red shift of spectral lines due to the gravitational potential should not exist, then the general theory of relativity will be untenable.' Sir Karl Popper, in his recent *Autobiography*, indicates that his own falsifiability criterion owed at its origin much to what he perceived to be Einstein's example, and he cites this specific sentence, which he says he read with profound effect when he was still in his teens.

Those of us who have admired Sir Karl's work can only be grateful that he came upon Einstein's sentence in the 1920 edition that helped set him on his path. In its earlier editions and frequent printings of 1917, 1918, and 1919, Einstein's book had ended very differently. There, Einstein acknowledged that his general relativity theory so far had only one observable consequence, the precession of the orbit of Mercury, whereas the predicted bending of light and of the red shift of spectral lines owing to the gravitational potential were too small to be then observed. Nevertheless, Einstein drew this conclusion, in a sentence which concluded his book in its first fifteen printings: 'I do not doubt at all that these consequences of the theory will also find their confirmation.'

Suspension of disbelief

To illustrate that Einstein is not so different from other scientists when it comes to the willingness to suspend disbelief, it will be worth making an excursion to watch how an experimentalist of great skill went about his business in much the same way, but in the privacy of his laboratory. Some time ago I came across the laboratory notebooks of R. A. Millikan, containing the raw data from which he derived his measured value of the basic unit of electric charge, the electron. Millikan's earlier attempts in this direction had been quite vulnerable, and had come under bitter attack from a group of research physicists at the University of Vienna, chiefly Felix Ehrenhaft, who believed not in a unitary but in a divisible electron, in subelectrons carrying charges such as one-fifth, one-tenth, or even less of the ordinary electron. Now, in gearing up his response in 1911-12, Millikan had two strong supports for his counter-attack. One was his unflagging preconception that there is only one 'electrical particle or atom', as he put it, a doctrine he believed to have been proposed first and convincingly by Benjamin Franklin. His other support was the kind of superb skill described in the passage quoted from Bernard DeVoto's book.

Millikan's publication came in the August 1913 issue of the *Physical*

Review, and effectively ended the scientific portion of the controversy. It contains data for 58 different oil drops on which he has measured the electric charge. He assures his readers, in italics: '*It is to be remarked, too, that this is not a selected group of drops, but represents all of the drops experimented on during 60 consecutive days.*' Four years later, in his book *The electron*, Millikan repeats this passage, and all the data from the 1913 paper, and he adds for extra emphasis: 'These [58] drops represent all of those studied for 60 consecutive days, no single one being omitted.'

At the Millikan Archive of the California Institute of Technology, the laboratory notebooks are kept from which the published data were derived. If we put our eye to that key-hole in the service of the ethology of science, we find there were really 140 identifiable runs, made over a period of six months, starting in October 1911. Anyone who has done research work in a laboratory cannot help but be impressed by the way Millikan handles his data, and by the power of a presupposition shrewdly used.

To prepare for the proof from Millikan's laboratory records, let me remind you of the chief point of Millikan's oil drop experiment. In a simplified form that nevertheless retains the scientific essentials as well as its beauty and ingenuity, it is now a standard exercise in the repertoire of school physics. A microscopic oil droplet is timed as it falls through a fixed distance in the view field. It will have some net electric charge to begin with, if only owing to the friction that acted on it when it was initially formed and expelled from the vaporizer. Other electric charges may be picked up from time to time as the droplet encounters ionized molecules in the gas through which it falls. Neither of these charges influences the droplet's motion, so long as it falls freely in the gravitational field. But when an electric field of the right sign and magnitude is suddenly applied, the drop will reverse its course, and will rise the more rapidly the larger the electric charge on it. Comparing the times taken for falling and subsequent rising allows one to calculate the net charge owing to friction on the droplet, q_{fri}, while comparing the times for alternate risings yields the net charge owing to the encounter with gas ions, q_{ion}.

As one watches the same droplet over a long time, through its many up and down excursions, one can accumulate a large number of values for q_{fri} and q_{ion}. Now the fundamental assumption Millikan makes throughout his work is that q_{fri} as well as q_{ion} are always some integral multiple of a unit charge equal in magnitude to the charge of the electron,

e. Conversely, from the full set of data, he can determine the magnitude of e which is common to all of the values obtained for q_{fri} and q_{ion}, both being assumed to be always equal to 1, or 2, or 3 . . . × e. These assumptions become plausible when the scatter of values for e turns out to be small when computed from either q_{fri} or q_{ion}—and when the mean values of e, so differently based, are nevertheless closely equal for a given droplet.

This is just what happens for the 58 'runs' or droplets discussed in the August 1913 paper of Millikan. One of the runs made on the Ides of March, 1912, and recorded in Millikan's laboratory notebook, is typical.[4] The difference between the values of e, computed on the two different bases, is only about 0.1 per cent, and not far from the limits set by the apparatus itself. The page on which both the data and the calculations appear records Millikan's exuberance and pleasure in the lower left corner: 'Beauty. *Publish* this surely, *beautiful*!'

Millikan continued immediately to take data on another oil droplet, entering the data on the next page. This time things did not go well. It was now a heavier drop, hence its time of fall was shorter. The numbers of charges it picked up as it went along were not greatly different, and it did not stay in view as long as one would have liked. Now the difference between the average values of e, calculated from q_{fri} and q_{ion} respectively, were 1 per cent apart, instead of 0.1 per cent. So Millikan notes in his private laboratory book on that page: '*Error high* will not use',—and indeed it does not appear among the 58 droplets that made it into the final paper. From Millikan's point of view, it was a failed run, or, in effect, no run at all. The magnitude of the difference in the values of e obtained in those two ways was awkwardly large, although not so surprising as to threaten Millikan's fundamental assumptions. Instead of wasting time, he simply went on to the next set of readings with another droplet.

But the discarded set of observations—and many others like it in the same laboratory notebook—would have appeared very differently if examined from another set of presuppositions. Thus, the entries make excellent sense if one assumes that the smallest charge involved in the oil drop experiment is not e, but, say, $1/10\,e$. In that case, the number of charges on a given droplet would not have been, in succession, 11, 13, and 14, as Millikan had to assume, but could have been 109, 129, and 139; and correspondingly, the difference between the (now smaller) elementary charges obtained in the two ways would be of the order of 0.1 per cent, instead of Millikan's 1 per cent. The 'high' error was the direct result of Millikan's assumption that the smallest charge in nature could not be a fraction of the charge of the electron e, as also determined (although more indirectly) by different methods in many other branches of physics.

Millikan's decisions seem to us now eminently sensible; but the chief point of the story is that, in 1912, Millikan's assumption of the unitary nature of the electric charge was by no means the only one that could be made. On the contrary, a chief reason for his work at the time was to perfect his method and support his claim against the constant onslaught of Felix Ehrenhaft and his associates who, for a couple of years, had been publishing experiments in support of their own, precisely opposite presupposition, namely in favour of the existence of *sub*-electrons.

It is also part of the historical setting that, at the time, Millikan was really just beginning belatedly on his career as a research physicist, whereas Ehrenhaft—at a venerable and much better equipped university—had begun to be widely recognized and rewarded years earlier as a fast-rising star in experimental physics. It was only after losing the argument with Millikan, and probably as a result of it, that he began a rapid decline as a scientist. When Millikan was doing his experiments, the matter was still in the balance. If Ehrenhaft had had access to Millikan's notebook, he would have found for his purposes precisely those runs most valuable which, for Millikan, were 'failed'.

Conversely, Millikan's own presupposition helped him to identify difficulties of the usual experimental nature which he did not feel were worth following up. For many of those he entered a plausibility argument on the spot (e.g. that the battery voltages must have changed, convection interfered, the stop-watch might be in error). The laboratory notebooks record Millikan's frank comments in such cases. The most revealing of the lot—revealing both of Millikan's insights that dust particles might intrude in the observation chamber, and of the willingness to take risks on behalf of his presupposition—is a marginal note entered for a long run that yielded a value of e far outside the expected limit of error: '$e = 4.98$ which means that this could not have been an oil drop.'

Like the trapper in Indian country, he was advancing on dangerous territory, but with a framework of beliefs and assumptions within which judgments are possible. The chief gain was the avoidance of costly interruptions and delays that would have been required to pin down the exact causes of discrepant observations. Obviously, this is not a method we recommend to our beginning students. But obviously also, any discussion of the advance of science that does not recognize the role of the suspension of disbelief at crucial points is not true to the activity.[5]

Towards a third mechanism

Einstein would not have been surprised by Millikan's notebook. Perhaps because of his experience with Kaufmann, he took a dim view of new experiments that, like Ehrenhaft's, made strong claims not explainable in terms of theoretical systems which embrace a greater complex of phenomena. Very early in his career, Einstein had, it seems to me, formed a clear view about the basic structure of nature: at the top there is a small number of eternal, general principles or laws by which nature operates. These are not easy to find–partly because God is subtle, and partly because they do not stop at the boundaries between fields that happen to be occupied by different theories.

Below this upper layer of a few grand laws lies a layer of experimental facts–not the latest news from the laboratory, but hard-won, well-established, aged-in-the-bottle results, many going back to Faraday and Fresnel, and now indubitable. These experiences or key phenomena are the necessary consequences of the visible compliance with the general laws.

But between these two solid levels is the uncertain and shifting region of concepts, theories, and recent findings. They deserve to be looked at, but sceptically; they are man-made, limited, fallible, and if necessary, disposable. Einstein's attitude was perhaps best expressed in a remark reported to me by one of his colleagues in Berlin, the physical chemist Herman F. Mark: 'Einstein once told me in the lab: "You make experiments and I make theories. Do you know the difference? A theory is something nobody believes except the person who made it, while an experiment is something everybody believes except the person who made it".'

What, then, must one conclude from Kaufmann's fatal predisposition for the ether; Max Planck's predisposition for the continuum and against discreteness; Robert Millikan's predisposition for a discrete rather than a divisible electron; Einstein's predisposition for a theory that encompasses a wide rather than a narrow range of phenomena–all in the face of clear and sometimes overwhelming difficulties? These cases–which can be matched and extended over and over again–show that some *third mechanism* is at work here, in addition to the phenomenic and analytical. And we can find it right in Einstein's lecture on the method of theoretical physics: the two-dimensional model in it, which first strikes the eye, gives way on closer examination to a more sophisticated and appropriate one. In addition to the two inseparable but antithetical components there is indeed a third–not as clearly articulated here as in some others of Einstein's essays, but present nevertheless. The arguments for it float above the plane bounded by the empirical and logical components of the theory.

Einstein launches on it by reminding his audience, as he often did, that the previously mentioned phenomenic-analytic dichotomy prevents the principles of a theory from being 'deduced from experience' by 'abstraction'–that is to say, by logical means. 'In the logical sense [the fundamental concepts and postulates of physics are] free inventions of the human mind', and in that sense different from the unalterable Kantian categories. He repeats more than once that the 'fundamentals of scientific theory' are of 'purely fictitious character'. As he puts it soon afterwards, in the essay 'Physics and reality' (1936), the relation between sense experience and concept 'is analogous not to that of soup to beef, but rather to that of check number to overcoat.' The essential arbitrariness of reference, Einstein explains in the Spencer Lecture, 'is perfectly evident from the fact that one can point to two essentially different foundations'–the general theory of relativity, and Newtonian physics–'both of which correspond with experience to a large extent'–namely, with much of mechanics. The elementary experiences do not provide a logical bridge to the basic concepts and postulates of mechanics. Rather, 'the axiomatic basis of theoretical physics . . . must be freely invented.'

But if this is true, an obvious and terrifying problem arises, and Einstein spells it out. He writes: How 'can we ever hope to find the right way? Nay, more, has this right way an existence outside our illusions? Can we hope to be guided safely by experience at all when there exist theories such as classical mechanics, which do justice to experience to a large extent, but without grasping the matter in a fundamental way?'

We have now left the earlier, confident portion of Einstein's lecture far behind. The question raises itself whether the activities of scientists can ever hope to be cumulative, or whether we must stagger from one fashion, conversion, or revolution to the next, in a kind of perpetual, senseless Brownian motion, without direction or *telos*.

At that point, Einstein issues a clarion call: 'I answer with full confidence that there is, in my opinion, a right way, and that we are capable of finding it.' Here, Einstein goes suddenly beyond his earlier categories of empirical and logical efficacy, and offers us a whole set of selection rules with which, as with a good map and compass, that 'right way' may be found. Here, there, everywhere, guiding concepts emerge and beckon from above the previously defined plane to point us on the right path.

The first directing principle Einstein mentions is his belief in the efficacy of formal structures: The 'creative principle resides in mathematics'—not, for example, in mechanical models. On the next page, there unfolds itself a veritable hymn to the guiding concept of simplicity. Einstein calls it 'the Principle of searching for the mathematically simplest concepts and their connections', and he cheers us on our way with many examples of how effective it has already proved to be: 'If I assume a Riemannian metric [in the four-dimensional continuum] and ask what are the *simplest* laws which such a metric can satisfy, I arrive at the relativistic theory of gravitation in empty space. If in that space I assume a vector field or anti-symmetrical tensor field which can be derived from it, and ask what are the simplest laws which such a field can satisfy, I arrive at Maxwell's equations for empty space'; and so on, collecting victories everywhere under the banner of simplicity.

And over there, at the bottom of another page, we find two other guiding concepts in tight embrace: the concept of parsimony, or economy, and that of unification. As science progresses, Einstein tells us, 'the logical edifice' is more and more 'unified', the 'smaller the number [is] of logically independent conceptual elements which are found necessary to support the whole structure.' Higher up on that same page, we encounter nothing less than 'the noblest aim of all theory', which is 'to make these irreducible elements as simple and as few in number as is possible, without having to renounce the adequate representation of any empirical content'.

Yet another guiding concept given in Einstein's lecture concerns the *continuum*, the field. From 1905 on, when the introduction of discontinuity in the form of the light quantum forced itself on Einstein as a 'heuristic' and therefore not fundamental point of view, he clung to the hope and programme to keep the continuum as a fundamental conception, and he defended it with enthusiasm in his correspondence. It was part of what he called his 'Maxwellian programme' to fashion a unified field theory. Atomistic discreteness and all it entails was not the solution but rather the problem. So here, in his 1933 lecture, he again considers the conception of 'the atomic structure of matter and energy' to be 'the great stumbling block for a unified field theory'.

One cannot, he thought, settle for this basic duality in nature, giving equal status both to the field and to its antithesis. Of course, neither logic nor experience forbade it. Yet, it was almost unthinkable. As he once wrote to his old friend, Michel Besso, 'I concede . . . that it is quite possible that physics might not, finally, be founded on the concept of field—that is to say, on continuous elements. But then out of my whole castle in the air—including the theory of gravitation and most of current physics—there would remain almost nothing.'

We have by no means come to the end of the list of presuppositions which guided Einstein. But it is worth pausing to note how plainly he seemed to have been aware of their operation in his scientific work. In this too he was rare. Sir Isaiah Berlin, in his book *Concepts and categories* [p. 159], remarked: 'The first step to the understanding of men is the bringing to consciousness of the model or models that dominate and penetrate their thought and action. Like all attempts to make men aware of the categories in which they think, it is a difficult and sometimes painful activity, likely to produce deeply disquieting results.' This is generally true; but it was not for Einstein. There are surely at least two reasons for that. It was, after all, Einstein who realized the 'arbitrary character' of what had for so long been accepted as 'the axiom of the absolute character of time, viz., of simultaneity [which] unrecognizedly was anchored in the unconscious', as he put it in his *Autobiographical notes*. 'Clearly to recognize this axiom and its arbitrary character really implies already the solution of the problem.' (Giving up an explicitly or implicitly held presupposition has indeed often had the characteristic of the great sacrificial act of modern science; we find in the writings of Kepler, Planck, Bohr, and Heisenberg that such an act is a climax of a period that in retrospect is characterized by the word 'despair'.)

Having recognized and overcome the negative, or enslaving, role of presuppositions, Einstein also saw their positive, emancipating potential. In one of his early essays on epistemology ('*Induction and deduction in physics*', 1919), he wrote: 'A quick look at the actual development teaches us that the great steps forward in scientific knowledge originated only to a small degree in this [inductive] manner. For if the researcher went about his work without any preconceived opinion, how should he be able at all to select out those facts from the immense abundance of the most complex experience, and just those which are simple enough to permit lawful connections and become evident?'

In essay after essay, Einstein tried to draw attention to this point of view, despite—or because of—the fact that he was making very few converts. The Herbert Spencer lecture can be seen as part of that mission. A decade and a half later, in his 'Reply to criticisms', we see him continuing in this vein. Thus, he acknowledges that the distinction between 'sense impressions' on the one hand, and 'mere ideas' on the other, is a basic conceptual tool for which he can adduce no convincing evidence. Yet, he needs this distinction. His solution is simply to announce, 'we regard the distinction as a category which we use in order that we might the better find our way in the world of immediate sensation.' As with other conceptual distinctions for which 'there is also no logical-philosophical justification', one has to accept it as 'the presupposition of every kind of physical thinking', mindful that 'the only justification lies in its usefulness. We are here concerned with "categories" or schemes of thought, the selection of which is, in principle, entirely open to us and whose qualification can only be judged by the degree to which its use contributes to making the totality [*sic*] of the contents of consciousness "intelligible".' Finally, he curtly dismisses an implied attack on these 'categories' or 'free conventions' with the remark that 'Thinking without the positing of categories and of concepts in general would be as impossible as is breathing in a vacuum.'

The thematic dimension

His remarkable self-consciousness concerning his fundamental presuppositions throughout his scientific and epistemological writings allows one to assemble a list of about ten chief presuppositions underlying Einstein's theory construction: primacy of formal (rather than materialistic or mechanistic) explanation; unity or unification; cosmological scale in the applicability of laws; logical parsimony and necessity; symmetry (as long as possible); simplicity; causality (in essentially the Newtonian sense); completeness and exhaustiveness; continuum; and of course constancy and invariance.

These ideas, to which Einstein was obstinately devoted, explain why he would continue his work in a given direction even when tests against experience were difficult or unavailable, or, conversely, why he refused to accept theories well supported by the phenomena but, as in the case of Bohr's quantum mechanics, based on presuppositions opposite to his own. Much the same can be said of most of the major scientists whom I have studied. Each has his own, sometimes idiosyncratic map of fundamental guiding notions—from Johannes Kepler to Steven Weinberg and his contemporaries.

With this finding, we must now re-examine the mnemonic device of the two-dimensional plane. I remove its insufficiency by defining a third axis, rising perpendicularly out of it. This is the dimension orthogonal to and not resolvable into the phenomenic or analytic axes. Along it are located those fundamental preconceptions, often stable, many widely shared, that show up in the motivation of the scientist's actual work, as well as in the end-product for which he strives. Since they are not directly derivable either from observation or from analytic ratiocination, they require a term of their own. I call them themata. While the scientist generally is not and need not be conscious of them, the historian of science can chart the growth of a given thema in the work of an individual scientist over time, and show its power upon his scientific imagination. Thematic analysis, then, is in the first instance the identification of the particular map of themata which, like the lines in a fingerprint, can characterize a scientist or a part of the scientific community at a given time.

Most of the themata are ancient and long lived; many come in opposing diads or triads that show up most strikingly during a conflict between individuals or groups that base their work on opposing themata. I have been impressed by the small number of thematic couples, or triads; perhaps fewer than 50 have sufficed us throughout the history of the physical sciences: and of course I have been interested to see that, cautiously, thematic analysis of the same sort has begun to be brought to bear on significant cases in other fields.[6]

With this conceptual tool we can return to some of the puzzles we mentioned earlier. Let me point out two. If, as Einstein claimed, the principles are indeed free inventions of the human mind, there should be an infinite set of possible axiom systems to which one could leap or cleave. Virtually every one of these would ordinarily be useless for constructing theories. How then could there be any hope of success, except by chance? The answer must be that the license implied in the leap to an axiom system of theoretical physics by the freely-inventing mind is the freedom to make such a leap, but not the freedom to make *any leap whatever*. The freedom is narrowly circumscribed by a scientist's particular set of themata that provide constraints shaping the style, direction, and rate of advance of the engagement on novel ground. And in so far as the individual maps of themata overlap, the so-called progress of the scientific community as a group is similarly constrained or directed. Otherwise, the inherently anarchic connotations of 'freedom' could indeed disperse the total effort. As Mendeleev wrote: 'Since the scientific world view changes drastically not only from one period to another but also from one person to another, it is an expression of creativity. . . . Each scientist endeavors to translate the world view of the school he belongs to into an indisputable principle of science.' However, in practice there is far more coherence than this implies, and we shall presently look more closely at the mechanism responsible for it.

A second puzzle was where the conceptual and even emotional support comes from which, for better or worse, stabilizes the individual scientist's risky speculations and confident suspensions of disbelief during the nascent phase. In case after case, as in the example of Millikan, we see that choices of this sort are made often on thematic grounds. Millikan was devoted to the atomistic view of electricity from the beginning, while his chief opponent, probably under the influence of Ernst Mach and his school, came to look for precisely the opposite evidence, e.g. subelectrons that in principle have no lower limit of charge at all. Similarly, Einstein and his opponents such as Kaufmann were divided sharply on the explanatory value of a plenum (ether), and on the range of fundamental laws across the separate branches of physics.

The Ionian enchantment

But of all the problems that invite attention with these tools, the most fruitful is a return visit to that mysterious place, early in Einstein's 1933 lecture, where he speaks of the need to pay 'special attention to the relations between the content of the theory and the totality of empirical fact (*Gesamtheit der Erfahrungstatsachen*).' The *totality* of empirical fact! It is a phrase that recurs in his writings, and indicates the sweep of his conscious ambition. But it does even more: it lays bare the most daring of all the themata of science, and points to the holistic drive behind 'scientific progress'.

Einstein explicitly and frankly hoped for a theory that would ultimately be utterly comprehensive and completely unified. This vision drove him on from the special to the general theory, and then to the unified field theory. In a letter to a biographer, Carl Seelig, Einstein likened his progress to the construction of an architectonic entity through three stages of development. Each stage is characterized by the adoption of a 'limiting principle', a formal condition which restricts the choice of possible theories. For example, in going from special to general relativity theory, Einstein had to accept, from 1912 on, that physical significance attaches not to the differentials of the space–time co-ordinates themselves, as the strict operationalists would insist, 'but only to the Riemannian metric corresponding to them'. This entailed Einstein's reluctant sacrifice of the primacy of direct sense perception in constructing a physically significant system; but otherwise he would have had to give up hope of finding unity at the base of physical theory.

The search for one grand architectonic structure is of course an ancient dream. At its worst, it has sometimes produced authoritarian visions which are as empty in science as their equivalent is dangerous in politics. At its best, it has propelled the drive to the various grand syntheses that rise above the more monotonous landscape of analytic science. This has been the case in the last decades in the physical sciences. Today's triumphant purveyors of the promise that all the forces of physics will eventually melt down to one, who in the titles of their publications casually use the term 'The Grand Unification', are in a real sense the successful children of those earliest synthesis-seekers of physical phenomena, the Ionian philosophers.

To be sure, as Sir Isaiah warned in *Concepts and categories*, there is the danger of a trap. He has christened it the 'Ionian Fallacy', defined as the search, from Aristotle to Bertrand Russell and our day, for the ultimate constituents of the world in some non-empirical sense. Superficially, the synthesis-seekers of physics, particularly in their monistic exhortations, appear to have fallen into that trap—from Copernicus, who confessed that the chief point of his work was to perceive nothing less than 'the form of the world and the certain commensurability of its parts', to Einstein's contemporaries such as Max Planck, who exclaimed in 1915 that 'physical research cannot rest so long as mechanics and electrodynamics have not been welded together with thermodynamics and heat radiation', to today's theorists who, in their more popular presentations, seem to imitate Thales himself and announce that all is ineffable quark.

A chief point in my view of science is that scientists, in so far as they are successful, are in practice rescued from the fallacy *by the multiplicity of their themata, a multiplicity which gives them the flexibility that an authoritarian research programme built on a single thema would lack*. I shall develop this, but I can also agree quickly that something like an Ionian Enchantment, the commitment to the theme of grand unification, was upon Einstein. Once alerted, we can find it in his work from the very beginning. In his first published paper (1901), he tries to understand the contrary-appearing forces of capillarity and gravitation, and exclaims in a letter to his friend Marcel Grossmann, 'It is a magnificant feeling to recognize the unity [*Einheitlichkeit*] of a complex of phenomena that to direct observation appear to be quite separate things'—such as capillarity and gravitation, the physics of micro- and macro- regions. In each of his next papers we find something of the same drive, which he later called 'my need to generalize'. He examines whether the laws of mechanics provide a sufficient foundation for the general theory of heat, and whether the fluctuation phenomena that turn up in statistical mechanics also explain the basic behaviour of light beams and their interference, the Brownian motion of microscropic particles in fluids, and even the fluctuation of electric charges in conductors. And in his deepest work of those early years, in special relativity theory, the most powerful propellant is Einstein's drive toward unification; his clear motivation is to find a more general point of view which would subsume the seemingly limited and contrary problems and methods of mechanics and of electrodynamics.

Following the same programme obstinately to the end of his life, he tried to bring together, as he had put it in 1920, 'the gravitational field and the electromagnetic field into a unified edifice', leaving 'the whole physics' as a 'closed system of thought'. In that longing for a unified world picture, a structure that encompasses 'the totality of empirical facts', one cannot help hearing the voice of Goethe's Faust who exclaimed that he longed 'to detect the inmost force that binds the world and guides its course'—or, for that matter, Newton himself, who wanted to build a unifying structure so tight that the most minute details would not escape it.

The unified *Weltbild* as 'supreme task'

In its modern form, the Ionian Enchantment, expressing itself in the search for a unifying world picture, is usually traced to Von Humboldt and Schleiermacher, Fichte and Schelling. The influence of the 'Nature Philosophers' on physicists such as Hans Christian Oersted—who in this way was directly led to the first experimental unification of electricity and magnetism—has been amply chronicled. At the end of the nineteenth century, in the Germany of Einstein's youth, the pursuit of a unified world picture as the scientist's highest task had become almost a cult activity. Looking on from his side of the Channel, J. T. Mertz exclaimed in 1904 that the lives of the continental thinkers are 'devoted to the realization of some great ideal. . . . The English man of science would reply that it is unsafe to trust exclusively to the guidance of a pure idea, that the ideality of German research has frequently been identical with unreality, that in no country has so much time and power been frittered away in following phantoms, and in systematizing empty notions, as in the Land of the Idea.'

Einstein himself could not easily have escaped being aware of these drives toward unification, even as a young person. For example, we know that as a boy he was given Ludwig Büchner's widely popular book *Kraft und Stoff* (*Energy and matter*), a book Einstein often recollected having read with great interest. The little volume does talk about energy and matter; but chiefly it is a late-Enlightenment polemic. Büchner comes out explicitly and enthusiastically in favour of an empirical, almost Lucretian scientific materialism, which its author calls a 'materialistic world view'. Through this world view, the author declares, one can attain 'the unity of energy and matter, and thereby banish forever the old dualism'.

But the books which Einstein himself credited as having been the most influential on him in his youth were Ernst Mach's *Theory of heat* and *Science of mechanics*. That author was motivated by the same Enlightenment animus, and employed the same language. In the *Science of mechanics*, Mach exclaims: 'Science cannot settle for a ready-made world view. It must work toward a future one. . . that will not come to us as a gift. We must earn it! [At the end there beckons] the idea of a unified world view, which is the only one consistent with the economy of a healthy spirit.'

Indeed, in the early years of this century, German scientists were thrashing about in a veritable flood of publications that called for the unification or reformation of the 'world picture' in the very title of their books or essays. Max Planck and Ernst Mach carried on a bitter battle, publishing essays directly in the *Physikalische Zeitschrift*, with titles such as 'The unity of the physical world picture'. Friedrich Adler, one of Einstein's close friends, wrote a book with the same title, attacking Planck. Max von Laue countered with an essay he called 'The physical world picture'. The applied scientist Aurel Stodola, Einstein's admired older colleague in Zurich, corresponded at length with Einstein on a book which finally appeared under the title *The world view of an engineer*. Similarly titled works were published by other collaborators and friends of Einstein, such as Ludwig Hopf and Philipp Frank.

Perhaps the most revealing document of this sort was the manifesto published in 1912 in the *Physikalische Zeitschrift* on behalf the new *Gesellschaft für positivistische Philosophie*, composed in 1911 at the height of the *Weltbild* battle between Mach and Planck. Its declared aim was nothing less than 'to develop a comprehensive *Weltanschauung*', and thereby 'to advance toward a noncontradictory, total conception [*Gesamtauffassung*]'. The document was signed by, among others, Ernst Mach, Josef Petzold, David Hilbert, Felix Klein, Georg Helm, Albert Einstein (only just becoming more widely known at the time), and that embattled builder of another world view, Sigmund Freud.

It was perhaps the first time that Einstein signed a manifesto of any sort. That it was not a casual act is clear from his subsequent, persistent recurrence to the same theme. His most telling essay was delivered in late 1918, possibly triggered in part by the publication of Oswald Spengler's *Decline of the west*, that polemic against what Spengler called 'the scientific world picture of the West'. Einstein took the occasion of a presentation he made in honour of Max Planck (in *Motiv des Forschens*) to lay out in detail the method of constructing a valid world picture. He insisted that it was not only possible to form for oneself 'a simplified world picture that permits an overview [*übersichtliches Bild der Welt*]', but that it was the scientist's 'supreme task'. Specifically, the world view of the theoretical physicist 'deserves its proud name *Weltbild*, because the general laws upon which the conceptual structure of theoretical physics is based can assert the claim that they are valid for any natural event whatsoever. . . . The supreme task of the physicist is therefore to seek those most universal elementary laws from which, by pure deduction, the *Weltbild* may be achieved.'

There is of course no doubt that Einstein's work during those years constituted great progress towards this self-appointed task. In the developing relativistic *Weltbild*, a huge portion of the world of events and processes was being subsumed in a four-dimensional structure which Minkowski in 1908 named simply *die Welt*—a Parmenidean crystal-universe, in which changes, e.g. motions, are largely suspended and, instead, the main themata are those of constancy and invariance, determinism, necessity, and completeness.

Typically, it was Einstein himself who knew best and recorded frequently the limitations of his work. Even as special relativity began to make converts, he announced that the solution was quite incomplete because it applied only to inertial systems and left out entirely the great puzzle of gravitation. Later he worked on removing the obstinate dualities, explaining for example that 'measuring rods and clocks would have to be represented as solutions of the basic equation . . . not, as it were, as theoretical self-sufficient entities'. This he called a 'sin' which 'one must not legalize'. The removal of the sin was part of the hoped-for perfection of the total programme, the achievement of a unified field theory in which 'the particles themselves would *everywhere* be describable as singularity-free solutions of the complete field-equations. Only then would the general theory of relativity be a *complete* theory.'[7] Therefore, the work of finding those most general elementary laws from which by pure deduction a single, consistent, and complete *Weltbild* can be won, had to continue.

There has always been a notable polarity in Einstein's thought with respect to the completeness of the world picture he was seeking. On the one hand he insisted from beginning to end that no single event, individually considered, must be allowed to escape from the final grand net. We noted that in the Herbert Spencer lecture of 1933 he is concerned with encompassing the 'totality of experience', and declared the supreme goal of theory to be 'the adequate representation of any content of experience' (translated in the first English version of the 1933 lecture, as delivered by Einstein, as 'the adequate representation of a single datum of experience'). He even goes beyond that; toward the end of his lecture he reiterates his old opposition to the Bohr–Born–Heisenberg view of quantum physics, and declares 'I still believe in the possibility of a model of reality, that is to say a theory, which shall represent the events themselves [*die Dinge selbst*] and not merely the probability of their occurence'. Writing three years later (*Physics and reality* 1936), he insists even more bluntly:

> But now, I ask, does any physicist whosoever really believe that we shall never be able to attain insight into these significant changes of single systems, their structure, and their causal connections, despite the fact that these individual events have been brought into such close proximity of experience, thanks to the marvellous inventions of the Wilson-Chamber and the Geiger counter? To believe this is, to be sure, logically possible without contradiction; but it is in such lively opposition to my scientific instinct that I cannot forego the search for a more complete mode of conception.

Yet, even while Einstein seemed anxious not to let a single event escape from the final *Weltbild*, he seems to have been strangely uninterested in nuclear phenomena, that lively branch of physics which began to command great attention precisely in the years Einstein started his own researches. He seems to have thought that these phenomena, in a relatively new and untried field, would not lead to the deeper truths. And one can well argue that he was right; not until the 1930s was there a reasonable theory of nuclear structure, and not until after the big accelerators were built were there adequate conceptions and equipment for the hard tests of the theories of nuclear forces.

Einstein's persistent pursuit of fundamental theory without including nuclear phenomena can be understood as a consequence of a suspension of disbelief of an extraordinary sort. It is ironic that, as it turned out, even while Einstein was trying to unify the two long-range forces (electromagnetism and gravitation), the nucleus was harbouring two additional fundamental forces, and moreover that after a period of neglect, the modern unification programme, two decades after Einstein's death, began to succeed in joining one of the nuclear (relatively short-range) forces with one of the relatively long-range forces (electromagnetism). In this respect, the labyrinth through which the physicists have been moving appears now to be less symmetrical than Einstein had thought it to be.

For this and similar reasons, few of today's working researchers consciously identify their drive towards the 'grand unification' with Einstein's. Their attention is attracted by the thematic differences, expressed for example by their willingness to accept a fundamentally probabilistic world. And yet the historian can see the profound continuity. Today, as in Einstein's time, and indeed that of his predecessors, the deepest aim of fundamental research is still to achieve one logically unified and parsimoniously constructed system of thought that will provide the conceptual comprehension, as complete as humanly possible, of the scientifically accessible sense experiences in their full diversity. This ambition embodies a *telos* of scientific work itself, and it has done so since the rise of science in the Western world. Most scientists, working on small fragments of the total structure, are as unselfconscious about their participation in that grand monistic task as they are about, say, their fundamental monotheistic assumption, carried centrally without having to be avowed believers. Indeed, Joseph Needham may well be right that the development of the concept of a unified natural science depended on the preparation of the ground through monotheism, so that one can understand more easily the reason that modern science rose in seventeenth-century Europe rather than, say, in China.

Thematic pluralism and the direction of advance

Difference between some themata and sharing of others: this formula in brief seems to me to answer the question why the preoccupation with the eventual achievement of one unified world picture did not lead physics to a totalitarian disaster, as an Ionian Fallacy by itself could well have done. At every step, each of the various world pictures in use was seen as a preliminary version, a premonition of the holy grail. Moreover, each of these various, hopeful but incomplete world pictures of the movement was not a seamless, unresolvable entity (unlike a 'paradigm'). Nor was each completely shared within a given sub-group. Each operated with a whole spectrum of separable themata, with some of the same themata present in portions of the spectrum in rival world pictures. Indeed, Einstein and Bohr agreed on far more than they disagreed. Moreover, most of the themata were not new—they very rarely are—but adopted from predecessor versions of the *Weltbild*, just as many of them would later be incorporated in subsequent versions of it. Einstein freely called his project a 'Maxwellian programme' in this sense.[8]

It is also for this reason that Einstein saw himself with characteristic clarity not at all as a revolutionary, as his friends and his enemies so readily did. He took every opportunity to stress his role as a member of an evolutionary chain. Even while he was working on relativity theory in 1905, he called it 'a modification' of the theory of space and time. Later, in the face of being acclaimed the revolutionary hero of the new science, he insisted, as in his King's College (1921) lecture: 'We have here no revolutionary act but the natural development of a line that can be traced through centuries.' Relativity theory, he held, 'provided a sort of completion of the mighty intellectual edifice of Maxwell and Lorentz'. Indeed he shared quite explicitly with Maxwell and Lorentz some fundamental presuppositions such as the need to describe reality in terms of continua (fields), even though he differed completely with respect to others, such as the role of a plenum.

On this model we can understand why scientists need not hold substantially the same set of beliefs, either in order to communicate meaningfully with one another in agreement or disagreement, or in order to contribute to cumulative improvement of the state of science. Their beliefs have considerable fine structure; and within that structure there is, on the one hand, generally sufficient stabilizing thematic overlap and agreement, and on the other hand sufficient warrant for intellectual freedom that can express itself it thematic disagreements. Innovations emerging from such a balance, even as 'far-reaching changes' as Einstein called the contributions of Maxwell, Faraday, and Hertz, require neither from the individual scientist nor from the scientific community the kind of complete and sudden reorientation implied in such currently fashionable language as revolution, Gestalt switch, discontinuity, incommensurability, conversion, etc. On the contrary, the innovations are coherent with the model of evolutionary scientific progress to which Einstein himself explicitly adhered, and which emerges also from the actual historical study of his scientific work.

Thus, I believe that generally major scientific advance can be understood in terms of an evolutionary process that involves battles over only a few but by no means all of the recurrent themata. The work of scientists, acting individually or as a group, seen synchronically or diachronically, is not constrained to the phenomenic-analytic plane alone, and hence is an enterprise whose saving pluralism resides in its many internal degrees of freedom. Therefore we can understand why scientific progress is often disorderly, but not catastrophic; why there are many errors and delusions, but not one great fallacy; and how mere human beings, confronting the seemingly endless, interlocking puzzles of the universe, can advance at all—even if not soon, or inevitably, to the Elysium of the single world conception that grasps the totality of phenomena.

Notes

1. P. Frank, *Einstein: his life and times*, p. 217. Knopf, New York (1947). As his correspondence with Frederick Lindemann (kept at the Einstein Archives in Princeton) shows, Einstein was 'particularly pleased' to enter into what he hoped would be 'regular contact' with Oxford, and he seems to have considered this lecture as part of that process. Indeed, Einstein added to the prefatory sentence cited above: 'May I say that the invitation makes me feel that the links between this University and myself are becoming professionally stronger?' At that time, Einstein had made up his mind not to return to Germany. But he had not yet decided, among various possibilities, where to settle.

2. It is of some importance to note here the publication history of Einstein's Herbert Spencer Lecture—a confusing history, although in that respect by no means different from that of many of Einstein's important essays. Einstein read his lecture in English, apparently the first time he had dared to do so at Oxford. As we know from his correspondence and diary of that time, he was studying English, but felt that he had a quite incomplete mastery of the language. The original manuscript of Einstein's lecture was in German, and has been

published in his collection *Mein Weltbild* pp. 113–19, Ullstein Verlag, Frankfurt am Main (1977), under the title 'Zur Methodik [*not* Methode] der theoretischen Physik'. In the English version, as actually delivered, Einstein acknowledged his 'thanks to my colleagues at Christ Church, Mr Ryle, Mr Page, and Dr Hurst, who helped me—and perhaps a few of you—by translating into the English the lecture which I wrote in German.'

Unfortunately, the English translation, as published as a small booklet by Oxford University Press (1933), left a good deal to be desired. Key portions of the original manuscript were rendered quite freely. Perhaps for this reason, a different English translation was prepared (by Sonja Bargmann) when Einstein later published a collection of his essays under the title *Ideas and opinions* pp. 270–6, Dell, New York (1954). In quoting from Einstein's Spencer Lecture, and indeed from his other publications, I have gone back to the corresponding original German essays and prepared my own translations where necessary.

3. P. Schilpp (ed.) *Albert Einstein, philosopher-scientist* p. 407. Open Court, Evanston, Illinois (1949).

4. I have given a detailed analysis of Millikan's work in Chapter 2 of my recent book, *The scientific imagination: case studies.* Cambridge University Press (1978).

5. Lest it be thought that Millikan was only lucky in guessing which of the data were really usable, I hasten to point out that he continued to exhibit his skill under much more difficult circumstances immediately after this work on the electron. He resumed his experiments on the photoelectric effect, for which he became best known. For ten years he worked with a wrong presupposition that light did not exhibit the quantization of energy. But in the end, he proved the quantum hypothesis experimentally—as he said in his Nobel Prize address, 'contrary to my own expectation'.

6. A brief survey of thematic analysis is provided in the Introduction and Chapter 1 of *The scientific imagination*, Cambridge University Press (1978).

7. A. Einstein, Autobiographical notes. In *Albert Einstein, philosopher-scientist* (ed. P. Schilpp) pp. 59–61, 81. Open Court, Evanston, Illinois (1949). Emphases in original.

In the Spencer Lecture, Einstein raises this whole problem only gently and at the end, by saying: 'Meanwhile the great stumbling block for a field theory of this kind lies in the conception of the atomic structure of matter and energy. For the theory is fundamentally non-atomic insofar as it operates exclusively with continuous functions of space', unlike classical mechanics which, by introducing as its most important element the material point, does justice to an atomic structure of matter. He does see a way out: 'For instance, to account for the atomic character of electricity the field equations need only lead to the following conclusion: The region of three-dimensional space at whose boundary electrical density vanishes everywhere always contains a total electrical charge whose size is represented by a whole number. In the continuum theory, atomic characteristics would be satisfactorily expressed by integral laws without localization of the entities which constitute the atomic structure.' In referring to the total electric charge whose size is represented by a whole number, he points of course to the result of R. A. Millikan's work.

8. The case is quite general. Thus, Kepler's world was constructed of three overlapping thematic structures, two ancient and one new: the universe as theological order, the universe as mathematical harmony, and the universe as physical machine. Newton's scientific world picture clearly retained animistic and theological elements. Lorentz's predominantly electromagnetic world view was really a mixture of Newtonian mechanics, as applied to point masses, determining the motion of electrons, and Maxwell's continuous-field physics. Ernest Rutherford, writing to his new protégé, Niels Bohr, on 20 March 1913, gently scolds him: 'Your ideas as to the mode of origin of spectra in hydrogen are very ingenious and seem to work out well: but the mixture of Planck's ideas [quantization] with the old mechanics make it very difficult to form a physical idea of what is the basis of it.' In fact, of course, Bohr's progress toward the new quantum mechanics via the correspondence principle was a conscious attempt to find his way stepwise from the classical basis.

I gladly express my indebtedness to Miss Helen Dukas and to the Estate of Albert Einstein for help and for permission to quote from Einstein's writings, and to the NSF and NEH for research support.

2 *The nature of the 'ultimate' explanation in physics*

ABDUS SALAM

International Centre for Theoretical Physics, Trieste and Imperial College of Science and Technology, London

Experiment alone can decide on truth . . . But the axiomatic basis of physics cannot be extracted from experiment.

Einstein: Herbert Spencer Lecture, June 1933

All science–physics in particular–is concerned with discovering *why* things happen as they do. The 'whys' so adduced must clearly be 'deeper', more universal, more axiomatic, less susceptible to direct experimental testing, than the immediate phenomena we seek to explain. And it is also well-known that it is the 'whys' of one generation which are often the points of departure for the next, to whom the earlier 'whys' can appear subjective, conditioned by 'unscientific' thinking, even wrong. The glory of science is that, this notwithstanding, we often arrive at correct predictions–at least to the extent of the experimental accuracies achievable and often better. I wish to speak about this continuing, ever-sharpening process about the 'whys' of physics in the context of the fundamental unification of physical forces which our generation is seeking for.

I can summarize my remarks in terms of three propositions:

1. The physics of the last century ascribed its deeper 'whys' to an all pervading mechanical ether. Einstein killed this ether, but he substituted for it something terribly close in spirit–a dynamical space-time manifold. Following Einstein, the deepest 'whys' of today's physics are to be found as manifestations of what we choose to assume as the basic attributes of the space-time manifold.
2. So far as dynamics is concerned, our final court of appeal, if all

else fails, is the 'bootstrap mechanism', the principle of self-consistency of the Universe. This principle may be traced back to the teleological dictum of Leibnitz—so savagely satirized by Voltaire in *Candide*—'The Universe is as it is for what else could it be.'

3. And finally, there are the 'laws of impotence'—so named by Max Born—which all 'whys' must respect. These laws of impotence—the glory of the physics of the twentieth century—consist of not-to-be-questioned admonitions like: thou shalt not conceive of velocities greater than that of light, to transmit signals; thou shalt quantize angular momentum in units of Planck's constant ($\hbar$).

There are other requirements governing the desirable 'whys', like economy of concepts and simplicity (Occam's razor), like 'naturalness', like beauty of the mathematics to be used (which somehow appears linked with its unreasonable efficacy). But these are well-known ideas and do not need elaboration.

To illustrate my remarks, and in particular the questioning by one generation of the 'whys' which led the generation before to (relative) truth, consider the classic example of the laws of planetary orbits and celestial gravity theory associated with the names of Kepler, Newton, and Einstein.

Kepler, the first person to give a quantitative description of laws of planetary motion, describes thus how he was led to their discovery.

> God reflected on the difference between the curved and the straight and preferred the nobility of the curved.
>
> Among bodies, omit . . . the irregular ones, and only retain those whose faces are equal in side and in angle. There remain five regular bodies of the Greeks: cube, pyramid, dodecahedron, icosahedron and octahedron. . . . If the five bodies be fitted into one another and if circles be described both inside and outside all of them, then we obtain precisely the number six of circles. . . . Copernicus has taken just six orbits of this kind, pairs of which are precisely related by the fact that those five bodies fit most perfectly into them.

Would this type of reasoning be considered 'scientific' today?†

Kepler described Copernicus as a 'blind man feeling his way with a staff'. It must have been this act of hubris which in turn had its nemesis in Köestler's description of Kepler as a 'sleep-walker'.

† Before we dismiss Kepler's reasoning, reflect on our own generation's partiality for the eightfold way, or for the *exceptional* Lie groups as candidates for symmetry groups in particle physics, stemming as this partiality usually does from the mathematical 'nobility' of these particular constructs!

Kepler was followed by Newton, who washed his hands of the entire search for 'why'; 'but hitherto I have not been able to discover the cause of . . . gravity from phenomena and I frame no hypotheses . . . Hypothesis . . . has no place in experimental philosophy.'

On this attitude of Newton, Einstein had this to say: 'We now realize with special clarity, how much in error are those theorists who believe that theory comes inductively from experiment. Even the great Newton could not free himself from this error (*Hypotheses non fingo*).'

But had Newton built no hypothesis into his gravity theory? According to Einstein, he had. This was the hypothesis that the gravitational charge (m) which occurs in Newton's Force Law ($F = m_1 m_2/r^2$) *exactly* equals inertial mass—the quantity of matter contained in the bodies which mutually attract. This is the so-called Equivalence Principle.

To see the force of Einstein's remark about Newton's assumption of the equality of gravitational charge with inertial mass, consider a hydrogen atom which consists of a proton and an electron. In making up the atom, the electron and the proton attract each other both electrically as well as gravitationally. The inertial mass of the atom equals proton's mass plus electron's mass minus the electrical, as well as the gravitational, binding energies. The ratio of the summed masses of the proton and the electron to the two varieties of binding energies is of the order of $1 : 10^{-8} : 10^{-47}$. Now, Eötvös (in the nineteenth century) in his celebrated torsion experiment, had in fact demonstrated that gravitational charge does equal inertial mass to the extent that, for the hydrogen atom for example, the electrical binding energy ($10^{-8} : 1$) contributes equally to both. But what about gravitational binding energy? Does the tiny relative number (10^{-47})—ascribable to gravitational binding—also affect inertial mass and gravitational charge equally? What would Newton say?

Einstein's own answer was unambiguous. His 'why' for the existence of the gravitational force ascribes this force to space-time dynamics, to the curvature of the four-dimensional space-time. His theory incorporates a 'strong equivalence' of gravitational charge with inertial mass. But there were rival theories—like those of Brans-Dicke's extension of Einstein's—which denied this equivalence so far as the gravitational binding energy is concerned. According to these theories, a part of this relative 10^{-47} would not show up in the gravitational charge.

The issue between Einstein and Brans-Dicke was joined, in March 1976, in two beautiful experiments, independently carried out by two teams; one led by Shapiro, the other by Dicke himself. These epic experiments consisted of measuring the mean (Kepler) positions of the earth and the moon to ± 30 cm through lunar laser ranging measurements. For heavenly test bodies as massive as these, the relative ratio of the gravitational binding energy to the total mass is in excess of 10^{-12} : 1 (and not the miserable, unmeasurable ratio 10^{-47} : 1 obtaining for the hydrogen atom).

To nobody's surprise—except perhaps to Dicke's—Einstein's strong equivalence principle proved to be correct. Dicke's own theory must be discarded, at least to all reasonable values of a new, adjustable parameter in his theory.†

To summarize, Kepler, Newton, and Einstein each started with a different 'why' for broadly the same set of phenomena. (To be more precise, Newton disclaimed any attempt at formulating a 'why' for gravity theory—even though he apparently did build into it an equivalence hypothesis, justified later by Einstein's totally different approach.) Each theory gave predictions commensurate and better than the accuracies of the experiments *then* possible. However, at present, Einstein's approach remains the deepest—and the most accurately predictive—that we know of for explaining the existence—the *raison* behind—one of Nature's fundamental forces (gravity). Will this for ever be the case? Will this theory need modifications, extensions, become part of a bigger whole; will it even have to be discarded altogether, together with all its axiomatic sub-structure?

Einstein believed that the discovery of the deep 'why', underlying the other forces of nature will also follow the pattern of 'geometrization' of gravity that he had given to physics. Before I consider this, let me take one more example of differing styles of the offered 'whys' at different epochs of physics. The example is from one of the other fundamental forces of Nature—electromagnetism. Maxwell, you may recall, predicted the existence of electromagnetic radiation on the basis of the 'displacement current' which he invented. This is one of the greatest feats of inventive discovery man has ever made—a discovery with few parallels in the change it brought about in the world we live in. Today an A-level student would demonstrate for you the necessity of a 'displacement current' from the conservation law of electric charge. But Maxwell himself went through a tortuous—and what today might

† Notice, like old soldiers, theories never die; they simply fade away. Thus, one could still save Brans-Dicke's theory, but only by assuming an outrageous 'unnatural' value for this adjustable parameter. Other phenomena would then be affected but they are (hitherto) untestable.

be considered an untenable—deduction based on a mechanical model of the ether. In Einstein's phrase, '[This] great change [was] brought about by Faraday, Maxwell, and Hertz—as a matter of fact, half-consciously, and against their will—[because] all three of them, throughout their lives considered themselves adherents of the mechanical theory [of the ether].' Notwithstanding this, does anyone dare feel superior to Maxwell? Even after what I just quoted from Einstein, listen to his reverence for Maxwell: 'Imagine Maxwell's feelings when the differential equations he had formulated proved to him that the electromagnetic fields spread in the form of polarized waves and with the speed of light. To few men in the world has such an experience been vouchsafed.'

Consider now the forces of electromagnetism, and the two nuclear forces, weak and strong, responsible for radioactivity phenomena and for fission and fusion respectively. Recently theory suggested and experiment confirmed that the weak nuclear force combines with electromagnetism—just as magnetism combined with electricity in the hands of Faraday and Maxwell a century ago—into one single, all embracing *electroweak* force. The secret of this unification† lay in the extension of the so-called gauge ideas (well-known in electromagnetism) to the weak nuclear force. The characteristic of a gauge force is that such forces are proportional to the 'charges' carried by the particles (e.g. $F = (e_1e_2)/r^2$ for electromagnetism, $F = (m_1m_2)/r^2$ for gravity, e_1 and e_2 are electric charges and m_1 and m_2 the gravitational charges (masses) carried by the two particles).

What has been shown is that analogous to the electric charge, there exists three weak charges which determine the strength of the weak nuclear force and that these three charges—together with the electric—form four components of a 'single' entity, each component transformable one into the other, through the operations of the group structure

† A crucial role in the demonstration of this electroweak unification was played by the ideas of 'spontaneous' symmetry breaking. To motivate these, one has to invoke self-consistency (my second proposition, see p. 29) and to build in a special type of symmetrical potential into the structure of the theory—a potential which (surprisingly enough) yields solutions with less symmetry than what we started from. This potential should guarantee that the weak nuclear force remains short-range as observed, without affecting the long-range character of the electromagnetic force. There is a welcome price which one pays for inventing such a potential; one predicts the existence of a hitherto undetected particle—the so-called Higgs particle—which is currently being searched for. This particle is welcome, for its existence would show that we are on the right track.

It is this sort of quantitative prediction, which distinguishes our use and our version of the self-consistency principle in physics, from empty philosophizing.

$SU(2) \times U(1)$ acting on an 'internal symmetry space'. I shall attempt to explain what I mean, more humanly, in a moment. But to complete the story: the future theoretical expectation is that the strong nuclear force is also a gauge force and the corresponding strong nuclear charges will eventually unite with the electroweak charges to make up a single entity, belonging to a still larger 'internal symmetry group', of which the electroweak $SU(2) \times U(1)$ is a part.† From the concept of the electroweak force we shall, we hope, progress soon to the concept of a unified *electro-nuclear* force, comprising electromagnetism as well as the two types of the nuclear force.

I have used the word 'internal symmetry space' to designate that mysterious something which provides the present 'why' for these unified gauge theories. Charge—electric, weak-nuclear, strong-nuclear—is a manifestation of the existence of an 'internal' symmetry structure and of the postulated symmetries of laws of physics for rotations and other transformation in this mysterious internal space. The analogy of the internal space is with the familiar space-time. And the analogy of the electric and nuclear charges is with the gravitational charge—the inertial mass—which is associated with the translation-symmetry of the four-dimensional space-time† continuum.

The question which arose in the 1930s when 'internal symmetry spaces' were first invented by Heisenberg and Kemmer and which has become more and more insistent with the success of gauge ideas is this: Are these 'internal spaces' purely mathematical constructs, or do they represent realistic adjuncts to the four-dimensional space-time manifold we are familiar with?

To take one example, one of the attempts currently being made is to describe physics in an eleven-dimensional space-time manifold. Of the eleven dimensions, four are the familiar space-time dimensions whose curvature is related to gravity and the other seven dimensions correspond to an internal symmetry space. In the theory advanced, the

† Experiments to demonstrate this are planned by the Indian–Japanese, Brookhaven–Irvine–Wisconsin and Milan–Turin–CERN collaborations. These are experiments designed to demonstrate that the proton is unstable with a half-life of the order of 10^{30} years. Hitherto the proton has been believed to be stable. (Compare 10^{30} years with the unmentionably tiny life of the Universe (of the order of 10^{10} years).)

† Translation-symmetry is the statement that the laws of physics are independent of the location of where an experiment to test them is performed. This is one example of symmetry which we choose to ascribe to space-time structure; cf. the first proposition p. 29. The experimental consequence of this assumed symmetry is the empirically testable conservation of energy and momentum.

seven dimensions curled in upon themselves 10^{-43} seconds after the Big Bang, attaining a size of the order of 10^{-33} cm and no more. We live on a cylinder in eleven-dimensional space, our major source of sensory apprehension of these extra dimensions being the existence of charges—electric, weak-nuclear, and strong-nuclear and the corresponding forces as manifestations of their curvature. Thus will Einstein's final dream (with which he lived for thirty-five years) of uniting gravity with the other (electronuclear) forces be eventually realized.

An exciting idea, which may or may not work quantitatively! But one question already arises; why the difference between the four familiar space-time dimensions and the seven internal ones? Why may the one lot curl in upon themselves, while the other lot do not? For the present, we shall make this plausible through the self-consistency principle; we shall invent a potential which will guarantee this as the only stable self-consistent dynamical system which can exist. There will be subtle physical consequences of this perhaps, in the form of remnants, like the black body radiation which was a remnant of the Big Bang. We shall search for these. Even if we find them, the next generation may question this entire mode of thought—particularly if a small discrepancy with our predictions is detected—and the cycle of questioning and answering might start all over again. Even today, an obvious question would be: Why eleven dimensions; why not a wholesome number, like thirteen? Or is this once again, due to the operation of the 'bootstrap', the self-consistency principle?

There is an alternative suggestion to these extra dimensions which seeks to explain charges (other than gravitational) within the context of no more than a conventional four dimensional space-time. This suggestion, due to Wheeler, Schemberg, and Hawking, does not add in new dimensions; it instead associates the electric and the nuclear charges to space-time topology—space-time Gruyere-cheesiness, wormholes of the granular size of the order of 10^{-33} cm. The idea is attractive. Topology, you may recall, is concerned with 'global' aspects as contrasted with the 'differential' aspects of the present tradition in physics. It thus represents a real break with the past. Unfortunately—and I say this deliberately and ungratefully, in order to provoke—my own feeling is that the mathematics of topology, in respect of what we need, has not progressed beyond the Möbius strip and the Klein-bottle. Topology, as a language for physics, is not yet capable of supporting the edifice the physicist may wish to erect on it. Could it be that our generation will be defeated by the lack of development of a necessary mathematical discipline in a direction that we need? This has never happened before in the history of physics, but on this note, I would like to leave you to ponder on the deeper 'whys' appropriate to the physics of today—and tomorrow.

3 *Rationality in science*

HENRY HARRIS

University of Oxford

One interesting consequence of the later work of Karl Popper is that the present generation of philosophers of science concerns itself with the sociology, history, or politics of science, but not with its logical structure. I think this state of affairs has arisen because Popper's criticisms have not only undermined the classical picture of the scientific method, based on the ideas of Bacon and Mill, they also eventually undermine his own position, so that nothing in the way of a coherent logical structure for science remains. For Thomas Kuhn, for example, the attempt to find such a structure is futile;[1] for Paul Feyerabend 'anything goes'.[2]

Popper's criticisms of the classical doctrines have their origin in dissatisfaction with the logical structure of induction. Popper calls this 'Hume's problem',[3] because some of the fundamental difficulties associated with the process of induction appear first to have been stated specifically by Hume in the *Treatise on human nature*.[4] Hume argues that there can be no possible *logical* justification for expectations about future events based on experience of past events: more specifically, that no matter how often one repeats an observation that appears to establish some form of concatenation between two events, one cannot from such repetition derive a logical structure that would justify the expectation that the same two events would be concatenated in any future observation. Popper accepts Hume's argument, and extrapolation from it brings him to the view that no proposition about empirical events can ever be definitively established or verified.[5] The impasse generated by this conclusion Popper circumvents by proposing that such propositions can, however, sometimes be definitively falsified.

They can be falsified, he claims, by tests that subject them, and the consequences that flow from them, to a critical screen of observation and experiment.

It seems to me that so long as some kind of formal distinction can be maintained between an empirical observation and a hypothesis, the idea that hypotheses can be falsified by observation and experiment, although far from novel, is at least plausible; and there remains much to be said for the view that scientific propositions can usually (although I suspect not always) be distinguished from non-scientific ones by the fact that they are, in principle, falsifiable. But Popper, in his later work, argues that all observations are made within a framework of preconceptions, so that all statements about observations necessarily embody a substrate of theory and hence necessarily have a hypothetical character. If this is so, then the critical screen of observation and experiment against which a hypothesis is to be tested itself becomes no more than a concatenation of hypotheses, all ultimately fallible. In that case we can never be sure that a hypothesis that has apparently been falsified is in fact false. All we can do is choose between various sets of unverifiable propositions, and there can be no logical rules for making the choice. If it is true that a hypothesis cannot be verified, then it is also true that it cannot in any final sense be falsified. In a more general context, Quine[6] makes a similar point when he argues that 'any statement can be held true, come what may, if we make drastic enough adjustments elsewhere in the system' and that 'conversely, by the same token, no statement is immune to revision'.

Popper's falsificationist doctrine thus leads eventually to the conclusion that no scientific proposition, at least none that rests on human observation, can ever be proved right or proved wrong. Experimental scientists find this difficult to accept. To make the conclusion more palatable, some people have, however, tried to provide methodological rules that would permit us to choose between rival hypotheses, even in this slippery world that permits neither proof nor disproof. Popper's own methodological rule is that one should prefer the hypothesis with the greater empirical content: a new hypothesis should be accepted only if it accounts for all the observations that were explained by the old hypothesis and if, in addition, it offers an explanation for some observation or observations that were not explained by the old hypothesis. On the face of it, this is an attractive notion, but it is of no use at all to the practising scientist. For, as Feyerabend[2] points out, the empirical content of a hypothesis cannot be known in advance; it unfolds as the hypothesis is explored. Few, if any, hypotheses account for all the observations that are known to have been made within their domain; and one hypothesis is often preferred to another because it sheds an interesting new light on one particular observation, even though it may explain other observations less well. The late Imre Lakatos,[7] though he wrote in the falsificationist tradition, abandoned the idea that rules can be found for falsifying (or, in the absence of falsification, for rejecting) any one hypothesis. The falsifiable unit is, according to him, not the hypothesis, but a series of interrelated hypotheses–the research programme. Research programmes, however, are not to be falsified by logical exercises; they are simply to be rejected when they cease to be productive. Lakatos defines progress in this context as 'the degree to which the series of theories leads us to the discovery of novel facts'. Regrettably, we are given no definition of what, in a falsificationist world, constitutes a fact, and hence no recipe for distinguishing a novel fact from a novel hypothesis. But even if we had been given such a recipe, it is difficult to see much more than hindsight in an analytical system that assesses research programmes simply in terms of their results; and it is perfectly clear that the most mistaken theories, or series of theories, may sometimes generate discoveries of the greatest importance. Need one do more than cite the discovery of America by Columbus?[8] It seems to me that all these attempts to rescue the pristine falsificationist position come very close to what Popper himself calls 'conventionalist stratagems': *ad hoc* changes of position devised to patch up a system that simply does not work.

If a scientific proposition can neither be proved nor disproved, what do scientists mean when they say that a proposition is right or wrong? Deciding whether propositions are right or wrong is, after all, a central part of organized science. The rewards that the scientific life offers to investigators whose propositions are usually right are very different from the rewards it offers to those whose propositions are usually wrong. Can it be that scientists operate within a framework of systematic delusions? This seems unlikely, for bridges do not often collapse under the weight of trains, electric lights usually turn on when one presses the light switch, susceptible bacterial infections are usually controlled if the appropriate antibiotic is given soon enough and in an adequate dose. It is difficult to escape the feeling that the scientist's view of the world must touch reality at least at some points.

Falsificationists are enormously impressed by human fallibility.

Indeed, their doctrine assumes that all scientific propositions are eventually found wanting. This pessimism has its origins, I believe, in the fact that many of those who now write about scientific method were originally physicists; and they support their arguments very largely with examples taken from the history of physics, especially the recent history of science are taken to be type cases. In his article 'Conjectural method,[9] to which Popper, Lakatos, Kuhn, and Feyerabend all contributed, the index has 32 entries under Einstein, 13 under Planck, and 16 under Bohr, but it makes no mention of Harvey, Mendel, or Pasteur. This preoccupation with the physics of the twentieth century has the consequence that situations that are quite exceptional in the history of science are taken to be type cases. In his article 'Conjectural knowledge',[10] Popper reveals that he first turned against the idea that there are 'established laws' because of Einstein's theory of gravity. 'There never was', he writes, 'a theory as well "*established*" as Newton's, and it is unlikely that there ever will be one; but whatever one may think about the status of Einstein's theory, it certainly taught us to look at Newton's as a "mere" hypothesis or conjecture.'

I do not see the effect of Einstein on Newton in quite this apocalyptic light. In their work on gravity, Newton and Einstein were concerned with highly abstract, and often intrinsically relativistic, concepts such as time, distance, velocity, mass, force, etc., and the contributions of both men in this field were essentially theoretical, that is, non-experimental. Few scientists operate in so ethereal a realm. Indeed, the dominance of theory over experiment in precisely this branch of physics is so atypical of science as a whole, that, in Germany during the period between the two wars, the question assumed important political dimensions.[11] In 1919 Einstein showed one of his students a telegram that had just arrived announcing the results of the English expedition that had gone to Principe to test Einstein's theory of relativity by measurements on an eclipse of the sun. The student was delighted at the good news, but Einstein appeared unmoved, saying that he had known that the theory was correct all along. When the student asked him what his attitude would have been if the results of the expedition had failed to confirm his theory, Einstein replied: 'Then I should have been sorry for the good Lord, for the theory is correct'.[12] I recall a meeting in which the late Jacques Monod, addressing an audience of biologists, put forward similar views about the relationship between theory and experiment; but that audience reacted with a mixture of amusement and outrage. The fact that Newton's unifying equations were replaced more than two centuries later by Einstein's more profound equations is certainly remarkable, but it is no argument at all for the notion that all scientific conclusions are similarly bound eventually to be displaced. I do not believe that it will ever be shown that the blood of animals does not circulate; that anthrax is not caused by a bacterium; that proteins are not chains of amino acids. Human beings may indeed make mistakes, but I see no merit in the idea that they can make nothing but mistakes.

There is a sense in which scientific propositions can be both verified and falsified, a sense in which scientific questions can be answered definitively. Let us suppose that at some point in pre-history, before the invention of water-craft, there lived in the village of Oxford a community that eked out a precarious existence by hunting and fishing; and that a similar community did the same thing in the village of Henley. Let us further suppose that the territory between the two villages was impassable because of the hostility of the intervening tribes and that this had always been so within the collective memory of the Oxfordians and the Henleyians. Let us suppose, finally, that there was none the less precarious communication between Oxford and Henley by a circuitous route leading through Aylesbury, so that each village knew something of the life of the others. One can envisage some Oxfordian, more inquisitive and perhaps more imaginative than his fellows, becoming engrossed, for one reason or another, in the question of whether the river that ran through Oxford was also the one that ran through Henley. The posing of the question—the realization that there is an important problem to be solved—is the critical first step in any scientific investigation, and often it is the one that calls for the greatest degree of originality. Many observers may see the same phenomenon, but it may be that one alone will see in it a problem so important that he commits to it the major part of his intellectual energy.

Having thus committed himself to his problem, our proto-Oxfordian must now adopt or devise techniques that might lead to its solution. He could, of course, merely sit down and think about it; but he would soon find, as modern scientists do, that you do not often get very far with a practical problem simply by taking thought. So he might begin by gathering information about the river at Henley from Henleyians who from time to time make the roundabout journey to Oxford. (Modern scientists would call this indirect evidence.) He might learn that the river at Henley had a different colour from the river at Oxford and that it ran more slowly. This might lead him to suppose that the two rivers were not the same; but he would clearly not regard this evidence as decisive. He might then put a number of logs daubed in a particular colour into the river at Oxford and ask some Henley contacts to let him know in due course whether these coloured logs reached Henley. If, after an appropriate lapse of time, he received the news that no daubed logs had been seen in Henley, he might feel a little more confident in the view that the river at Oxford did not run through Henley. But in discussing this view with some other Oxfordians who had now become interested in the problem, he would soon learn that they were less impressed by the fact that his logs had not been seen in Henley (what modern scientists call a negative finding), than they would have been if the logs had been seen there (a positive finding). His colleagues would immediately offer him a range of alternative explanations for his negative finding, and he would be obliged to go back and try harder. He now makes a major technical innovation. For the first time in the history of Oxfordian technology, he builds a raft large enough to carry a man; and, travelling under cover of darkness and hiding during the day in the shelter of trees along the river bank, he eventually makes his way down river to Henley. When he returns to Oxford his story is at first greeted with disbelief; but, after two or three of his colleagues have repeated the experiment, the idea that Oxford and Henley are joined by the one river is accepted. In due course, when the difficulties with the intervening tribes are finally overcome, the waterway becomes a normal avenue of communication and trade between the two villages.

It seems to me that the question whether the river that runs through Oxford also runs through Henley has been answered by our proto-Oxfordian's research, and answered definitively. The only escape from this conclusion is to argue that the investigator was in some way hallucinated and that this hallucination was shared by all his successors right down to the present day; but, as I have already pointed out, the proposition that investigators reach agreement about phenomena in the natural world because they share systematic delusions does not have much to commend it. One could, of course, still ask whether all the water that flows through Oxford reaches Henley; whether the river at Henley receives water from any other source; whether there are climatic or geological conditions that might, under certain circumstances, alter the course of the river. But these are different, and subsidiary, questions; and the fact that they might as yet be unresolved does not render inconclusive the answer that our proto-Oxfordian's research has given to the initial question. Well, you might say, this little anecdote does indeed show that at least in ancient days Oxford did contain one inhabitant with common sense; but you might fail to see any close analogy between primitive river exploration and the sophisticated armamentarium of modern science. Let us see whether we might not be able to bring the two a little closer together.

Take Harvey and the circulation of the blood.[13] The Galenical doctrine envisaged that the blood was concocted from food in the stomach and further decocted, to remove impurities, in the liver. From the liver a continuous ebb and flow was supposed to take place to all other parts of the body via the veins; and from the heart a similar ebb and flow was supposed to take place via the arteries. The exhausted venous blood was thought to be revived in the heart by admixture with arterial blood, which was said to receive a continuous supply of vital spirit from the air by way of the lungs and the pulmonary vein. Although in the seventeenth century the College of Physicians of London still fined its members if their lectures defected from this doctrine, the Galenical map of blood flow in the body had already encountered some difficulties before Harvey turned his attention to the problem. To begin with, anatomists had great difficulty in seeing how the arterial and venous blood were mixed together. They were unable to provide convincing evidence for the existence of passages through the septum that divided the two sides of the heart. Moreover, Realdus Columbus[14] had demonstrated (1) that the pulmonary vein contained blood and not air and (2), that the valves in the heart permitted the blood to flow in only one direction.

Harvey tested the observations of Columbus on the heart valves with the greatest care and confirmed that the blood could flow only from the heart into the arteries, but not in the reverse direction. He then turned his attention to the valves in the veins of the limbs and established that these permitted flow from the extremities to the heart, but, again, not in the reverse direction. The valves in the heart and blood vessels thus ensured that the flow of blood in the body was a one-way system. The significance of this conclusion now hinged on what estimate one made of the amount of blood that was pumped by the heart into the arterial system. Here Harvey introduced a methodological innovation. By extending his studies to cold-blooded animals, in which the heart beats much more slowly than in man, he was able to disentangle the complex series of muscular contractions that form the heartbeat and to obtain an acceptable measure of the cardiac output. The amount of blood pumped by the heart turned out to be vastly in excess of the amount that could be formed from food in the stomach and vastly in excess of the amount that could be consumed in the nutrition of the tissues. Harvey therefore concluded that the blood pumped by the heart into the arteries must be brought back to the heart by the veins: the blood must circulate. Of course, the discovery of the circulation of the blood left much unanswered. Did all the blood circulate? By what route did it pass from the arteries to the veins? Were there creatures in which the vascular system was arranged differently? But, again, these are different questions, and the fact that they were not resolved by Harvey's experiments does not undermine the finality of his principal conclusion. As in the case of our proto-Oxfordian's river exploration, Harvey's conclusion did not immediately find acceptance, as the exchange with Riolan amply demonstrates.[15] But Harvey's experiments were done again by other investigators, and the circulation of the blood eventually became, not a hypothesis, but a fact of life. In due course, the fundamental reorientation that Harvey produced in our understanding of the physiology of the human body had its practical consequences in medical therapy and in countless other areas of human enterprise.

An interesting piece of medical history, you might say, but still very remote from contemporary experimental science. Not at all. There is in the blood and lymph of vertebrates a small cell known as the lymphocyte, which is of crucial importance in the immunological responses of animals to foreign substances. Just before the Second World War, several physiologists had made estimates of the numbers of these cells that entered the blood from the lymph, and these numbers were found to be very large. The question that physiologists then posed was, in principle, exactly the same as the one posed by Harvey for the blood pumped by the heart into the arteries. Where did these large numbers of lymphocytes go? In the 1950s, J. L. Gowans, adopting new surgical techniques that permitted long-term cannulation of a major lymphatic trunk, was able to show that the number of lymphocytes entering the blood from the lymph under normal conditions was such that more than ten times the total number of lymphocytes present in the blood would have entered the bloodstream within one day. Either these cells were rapidly consumed in some way or they must, like the blood, circulate. By adopting new techniques that were made available by the wartime development of radioactive isotopes, Gowans was able to tag the lymphocytes and show that, indeed, they passed from the blood into the lymph and back again into the blood. As usual, the new idea met with some initial resistance;[16] but the experiments were soon done by others and, in a very few years, the recirculation of lymphocytes became recognized as a fact. I do not believe that this fact will one day turn out to be a mistaken hypothesis; and it has already begun to find practical application. Harvey's experiments and those of Gowans, like the river exploration of our proto-Oxfordian, do not replace one hypothesis by another in an unending series. Nor do they contribute to a process in which we come progressively nearer to the truth, but never quite attain it. These experiments provide definitive answers to the questions posed; they provide facts. Of course, not all scientific investigations have this fortunate outcome. There are areas of enquiry, for example explorations of the fundamental nature of matter or of the origin of the universe, in which an asymptotic approach to the truth does seem an apt metaphor to describe what is going on. But there are rather few such areas and they are exceptional in being inaccessible, or accessible only with great difficulty, to direct experimental intervention. And even in these areas, I see no compelling reason for assuming that the truth can *never* be attained.

I should now like to return to what Popper calls Hume's problem—the problem of induction. In all three of my historical anecdotes, the acceptance of the new doctrine required a repetition of the experiment by others. Scientists set great store by the repeatability of experiments. An investigator whose experimental work is consistently unrepeatable soon loses his credibility and his reputation. If Hume is right in his contention that no amount of repetition of an experiment can ever give logical grounds for predictions about the outcome of any future experiment, then what can scientists possibly be about when they react so fiercely to an experiment not being repeatable? Once again, I see no great future in the idea that scientists are peculiarly liable to systematic delusions. The answer, it seems to me, is that what scientists actually do when they repeat experiments is something very different from what Hume had in mind. *Sensu stricto*, of course, no experiment can ever be exactly repeated. Even if it were possible to do no more than alter the time and space co-ordinates of an observation, this alteration alone might introduce variables that affect what is actually observed. The credence that scientists give to an observation is not determined by the number of times that a particular investigator claims to have made it. A research student who sought to make his case simply by doing a particular experiment over and over again would soon be encouraged to seek employment elsewhere. When a scientist 'repeats' an experiment, what he is actually trying to achieve is some further understanding of the variables that commonly affect the outcome. He deliberately introduces minor modifications into his procedure in order to elicit further information. He kicks the problem around. What he hopes is that, by doing this, he will eventually gain enough information about the immediately relevant variables to enable him to write a set of specifications for other investigators. Of course, further experiments may uncover new, and often unsuspected, variables that prove to be relevant. However, if an investigator has thoroughly explored the variables that commonly affect the outcome of an experiment, and has written a good set of specifications, then his colleagues will usually be able to do the experiment by following these specifications. What scientists mean when they say an observation is 'unrepeatable' is that the specifications provided for making that observation are inadequate. Scientists would doubtless share Hume's view that the future is inevitably uncertain; but this does not prevent them from writing good specifications for building bridges, or, for that matter, for putting a man on the moon. That this is so indicates that there must be fundamental regularities in the natural world, even if only statistical regularities. If there were not, we could never know anything, and life could never have evolved.

This brings me to the question of 'laws' in science. Scientific laws are not, in general, universal statements: they do not claim that there are no exceptions in nature to the concatenations of events that they describe. There are indeed a few, essentially analytical, propositions, notably in the field of physics, that do perhaps have this character: for example, Newton's second law which defines force in terms of acceleration; but most scientific laws do not embody claims to universal validity. Any scientific proposition can be called a law if the variables that influence the observation in question have been adequately defined and if the observation has found a wide enough application. The term 'law' does not delineate a new category of scientific proposition; it is essentially an accolade that scientists give to propositions of very high quality. If the proposition that the blood circulates in animals were to be called Harvey's Law, few scientists would object; nor would they wish to withdraw the accolade because this turns out not to be true for some organisms that by other criteria can be classed as animals. Boyle's Law remains a law even though we now know that it is a special case applicable only to so-called 'perfect' gases. But one would much more readily call Harvey's discovery a law than, for example, the discovery that the red blood cells of camels are ellipsoidal. This is not, however, because the two propositions belong to different logical categories, or because there is a greater likelihood of finding a camel with non-ellipsoidal red cells than an animal in which the blood does not circulate; it is because Harvey's discovery has much greater explanatory power. Without taking for granted the circulation of the blood, large central areas of animal physiology and biochemistry would be incomprehensible, whereas the biological significance of the shape of camel red blood cells is for the present obscure. This does not mean that the study of ellipsoidal red cells might not eventually yield propositions that we should be glad to call laws. But the route from the particular to the general does not pass through mindless repetition of the same observation; it passes through an imaginative exploration of the ramifications of the phenomenon being observed.

Finally I come to the question of what constitutes rationality in science. The practising scientist accepts intuitively that there is a real world that exists independently of his own interaction with it and that this world has a coherent structure which determines the kind of interactions that are possible. He accepts that the analysis of such interactions can yield information about the structure of the world and that the information obtained in this way can be corrected and made progressively more accurate by the analysis of further interactions. He has no doubt that there is a distinction to be made between facts and theories even if it sometimes proves very difficult to determine what the facts are. Since scientists are remarkably successful in their activities, it would seem not unreasonable to suppose that these straightforward attitudes are not simply a reflection of philosophical naïveté (although scientists often are philosophically naïve), but that they stem from modes of operation that have proved fruitful. I do not think that one can hope to understand what these modes of operation are unless one appreciates that science, however formal its symbolism may sometimes become, is not an exercise in logic. When some philosophers talk about the logic of scientific investigation, to say nothing of the logic of discovery, I can only suppose that they speak metaphorically; for science is no more a matter of avoiding logical errors than the composition of music is a matter of avoiding consecutive fifths and octaves. If philosophers wish to define the general principles that underlie the practice of science, they must refrain from imposing artificial logical constructs on this highly complex human activity and must turn their attention to the detailed study of what scientists actually do. In this Thomas Kuhn[17] is undoubtedly right.

Science is a component of the evolutionary process by which human communities explore their environment and adapt to it. The observations that scientists make are made with sense organs and brains which are the end-products of evolutionary interactions between one particular animal species and one particular world. These organs register biologically relevant stimuli emanating from this particular world and, when they are functioning properly, they ensure effective interaction with what is sensed. Stimuli that are biologically relevant to other animal species, but not to man, may not be registered by human sense organs; and one can surmise that the whole human apparatus would be remarkably ill-adapted to interaction with a totally different world. The observations that we can make and the thoughts that we can have about them are thus intrinsically restricted by our evolutionary history. The classical epistemological question, how can I know what X observes when he looks at Y, finds its biological answer in the fact that the apparatus with which X observes and the apparatus with which I observe have a common evolutionary history. Both are determined by sets of genes with identical or closely similar nucleotide sequences. Both are severely limited in the stimuli that they are capable of registering and severely limited also in the ways in which these stimuli are processed and interpreted by the brain. When X and I look at Y, evolution has ensured that, within certain limits, we observe much the same thing. Variations are, of course, permitted (indeed, they are essential for further evolutionary development to occur), but only within a very narrow range; a sensory system or a nervous system that consistently registered stimuli in a highly aberrant way would render its possessor incapable of interacting effectively with the world about him. It is these restrictions, imposed on us by our evolutionary history, that provide us with the essential framework for checking whether our observations are right or wrong. We may agree with Quine when he argues that any statement may be held true if we make drastic enough adjustments elsewhere in the system, but, in our interactions with the real world, we are not free to make whatever adjustments we please.

Like all evolutionary processes, science operates essentially by trial and error. Scientists cannot determine whether hypotheses are right or wrong except by testing the predictions that they make against the constraints imposed by the real world. There are no intrinsic properties of a hypothesis, that is, no properties observable before the hypothesis is tested, which will permit scientists to recognize whether it is right or wrong. Yet scientists constantly decide to explore some hypotheses and decline to explore others. How is the choice made? Quine and Ullian[18] argue that there are properties, some apparently intrinsic in the sense that I have defined, which, if they cannot establish the validity of hypotheses, can at least make them more plausible. They give the following list: (1) conservatism (disturbing existing beliefs as little as possible); (2) generality; (3) simplicity; (4) refutability; and (5) modesty. A glance at the history of biology makes pretty short work of this list. The hypotheses of Harvey, Mendel, and Pasteur were anything but conservative; the structures of proteins and the mechanisms by which they are synthesized are anything but simple; and it is difficult to conceive of any observation on either present or past forms of life that would actually refute the theory of natural selection. (The scientific status of this theory is interesting. Despite the difficulty of specifying any set of conditions that would falsify it, it remains one of the greatest of scientific hypotheses, mainly, I believe, because of its remarkable explanatory power.) The important question, however, is not whether Quine and Ullian's list is right, but whether any list of intrinsic properties can confer plausibility. I doubt it. If there were a group of intrinsic properties which, either singly or in concert, made a hypothesis plausible, one would expect to find pretty general agreement among scientists about which hypotheses were worth exploring and which were not. What actually happens is very different. A new hypothesis almost invariably divides the relevant section of the scientific community; and when there are several competing hypotheses, it is usually found that each manages to attract some adherents who think it worth exploring. If there are intrinsic properties that confer plausibility on hypotheses, scientists on the whole don't seem to be able to recognize them.

A scientist seeking to assess whether a hypothesis is plausible first explores it in his head. Each scientist has his own mental store of facts, theories, and associations, and a private set of value judgements about the relative importance of the different elements in that mental store. He assesses the plausibility of a hypothesis by testing its predictions, as far as he can, against the constraints imposed by the private world of his imagination. Because this private world is necessarily an incomplete and imperfect representation of the real world and because the constraints imposed by the imagination are less stringent than those imposed by reality, individual scientists may come to very different conclusions about the plausibility of the same hypothesis. Yet we see that, in choosing their hypotheses, some scientists are consistently more fortunate than most of their colleagues. This is not, of course, a matter of luck, but a reflection of the fact that the imaginary worlds in which such scientists probe their hypotheses bear a closer relationship to reality than the imaginary worlds of their less fortunate colleagues. Scientists with a 'flair' for picking the right hypothesis simply have a more profound grasp of their subject.

When one speaks of the predictions that a hypothesis makes one uses the word prediction in a rather special sense: predictions, in this context, are analytic expansions of the hypothesis that lead the informed observer to expect that, if things are as the hypothesis proposes, certain kinds of interaction with the real world would be possible. The predictions are tested by seeing whether these interactions are in fact possible. A prediction may be so specific that, for the scientist, its fulfilment constitutes a definitive verification of the hypothesis in question. Consider again the hypothesis that the river that runs through Oxford also runs through Henley. This hypothesis predicts that one should be able to reach Henley by following the river from Oxford. If this turns out in fact to be possible, then, for the scientist, the hypothesis is definitively verified; and, *ipso facto*, the contrary hypothesis, that there is a different river running through each of the two towns, is definitively falsified. Some philosophers may throw up their hands at this point and declare that neither the truth nor the falsity of any hypothesis is logically entailed by any set of observations. To this I would answer that science is not logic: the conclusions that scientists draw from their observations are imposed not by the rules of logical entailment but by the operational rules laid down by man's evolutionary history. Not all scientific predictions are, however, as decisive as those that can be made about rivers running through Oxford and Henley. Some may, for technical reasons, be difficult to test or may be testable only by indirect methods which are subject to uncertainties of interpretation. A prediction may not be exclusive enough: it may be fulfilled by a set of observations, but these observations may be compatible with more than one hypothesis. A prediction may be complex in the sense that several different tests must be made to fulfil it, so that it may be fulfilled only in part. Naturally, scientists prefer, and seek to formulate, hypotheses that make decisive predictions, but, more often than not, they must make do with more open-ended formulations. This does not much matter in the long run, because the exploration of open-ended hypotheses may none the less yield useful information, and the imperfect models of the world that such hypotheses reflect may, by trial and error, be progressively improved. In science, as elsewhere, the proof of the pudding is not in the recipe, but in the eating.

Error is an integral part of science, for unfulfilled predictions, by indicating how the world is not, help, by elimination, to define how the world is. None the less, science, as a social activity, does not adopt a neutral attitude to error. The essential function of scientific organizations is to devise and promote communal procedures for minimizing and correcting error. This aim is entirely rational. For the fulfilment of a prediction (being right) at once opens the way to effective new interactions with the real world; whereas the failure of a prediction to be fulfilled (being wrong) in itself leads nowhere. In my three historical anecdotes I made a point of the fact that in each case the discovery eventually led to a change in practice. I hope you will not conclude from this that I adhere to the primitive notion that a scientific idea is to be judged simply by whether it is immediately useful or not. None the less, new insights into the real world have a way of being exploited sooner or later; and, even if the exploitation is not deliberate, it is in the nature of the evolutionary process that as our view of the world changes so our interactions with it change also. For these reasons scientists set much greater store by observations that demonstrate that something is the case (the predictions are fulfilled) than by those that demonstrate that something is not the case (the predictions are not fulfilled); and the social structure of science takes this into account in its system of rewards. If it did not do so, scientists might well be content to while away their time chasing innumerable will-o'-the-wisps.

Organized science sets its face against error in two other ways. It imposes a certain discipline in the presentation of observations and it provides a system of communal arbitration. Scientific publications are not merely a convenient method of disseminating information. Their editors insist that claims made be supported by the presentation of evidence and that the methods by which this evidence is obtained be reported in sufficient detail to permit other scientists, if they so wish, to make the observations themselves—a process that I have already referred to as the provision of a good set of specifications. Since the credibility and reputation of scientists hinge on the quality of the specifications that they provide, this publication convention induces them to take great pains to probe their own hypotheses very thoroughly before submitting them to public scrutiny. Communal arbitration flows naturally from publication. The publication of an interesting hypothesis attracts the attention of other scientists and induces them to devise their own tests for it. With many hands contributing, the testing process thus becomes wider, quicker, and more thorough. Communal arbitration is not, of course, infallible, but it has, on the whole, proved to be a remarkably efficient procedure for sorting out propositions that are right from those that are wrong.

In the light of all this, how does our rational scientist now look? Here is a thumb-nail sketch. He is a thoroughgoing empiricist who never troubles his head about the logic of what he is doing, but has no doubt that his activities yield information about the real world. He knows that he makes mistakes, but he also knows that he sometimes gets things right. He has no reservations about the ability of scientific procedures to verify and to falsify scientific propositions. He does his best to pick fruitful hypotheses to investigate and he goes to enormous trouble to test his ideas before he makes them public. He publishes his work in a form that permits other scientists to check it, and, although it may often go against the grain, he will accept the verdict of his colleagues in the end. But let there be no mistake. Rationality helps, but it is not a prescription for making discoveries. What makes one man see a problem where others see none, what makes one man see a new way of approaching a problem that others regard as intractable, these are different questions altogether.

Acknowledgement

I thank Sir Isaiah Berlin, Mr Paul Foulkes, and Mr Michael Shorter for forcing me to think a little harder about some of these problems.

Notes

1. T.S. Kuhn, Logic of discovery or psychology of research. In *Criticism and the growth of knowledge* (eds. I. Lakatos and A. Musgrave) pp. 1–23. Cambridge University Press (1970).
2. P. Feyerabend, *Against method.* Verso, London (1978).
3. According to Popper, the importance of 'Hume's problem' was first recognized in his article, Ein Kriterium des empirischen Charakters theoretischer Systeme. *Erkenntnis* **3**, 426 (1933).
4. In the 1874 edition of Hume's *A treatise on human nature* (eds. T.H. Green and T.H. Grose), Longmans, Green & Co., London, the relevant passages are to be found in Part III, Section 6, p. 388 and Section 8, p. 399.

5. K. R. Popper *passim*, but see especially the article, Conjectural knowledge, in *Objective knowledge* pp. 1–31, Clarendon Press, Oxford (1972).

6. W. V. Quine *From a logical point of view* Part II, Section 5, p. 43. Harvard University Press, Cambridge, Mass. (1953).

7. I. Lakatos, Falsification and the methodology of research programmes. In *Criticism and the growth of knowledge* (eds. I. Lakatos and A. Musgrave) pp. 91–195. Cambridge University Press (1970).

8. There is an excellent discussion of Columbus's geographical ideas in S. E. Morison *Admiral of the ocean sea: a life of Christopher Columbus*. Little, Brown & Co., Boston, Mass. (1942).

9. *Criticism and the growth of knowledge* (eds. I. Lakatos and A. Musgrave). Cambridge University Press (1970).

10. K. R. Popper *Objective knowledge* p. 9. Clarendon Press, Oxford (1972).

11. The politics surrounding theoretical physics in Germany during this period are discussed in A. D. Beyerchen, *Scientists under Hitler*. Yale University Press, New Haven (1977).

12. This story is taken from G. Holton, Mach, Einstein and the search for reality. *Daedalus* **97**, 653 (1968).

13. I have used as my source the excellent translation of Harvey's *De motu cordis* by G. Whitteridge. Blackwell Scientific Publications, Oxford (1970).

14. The contributions of Realdus Columbus are considered in detail in G. Whitteridge *William Harvey and the circulation of the blood*. Macdonald, London (1971).

15. Ibid., pp. 175–200.

16. See the discussion following the paper by Gowans *et al.*, and the paper by Yoffey *et al.*, in *Biological activity of the leucocyte* (eds. G. E. W. Wolstenholme and M. O'Connor) pp. 40–59, Ciba Foundation Study Group. Churchill, London (1961).

17. T. S. Kuhn, *The structure of scientific revolutions* (2nd edn). University of Chicago Press, Chicago (1970).

18. W. V. Quine and J. S. Ullian, *The web of belief* pp. 42–53. Random House, New York (1970).

4 True believers: the intentional strategy and why it works

DANIEL C. DENNETT

Department of Philosophy, Tufts University

DEATH SPEAKS

There was a merchant in Baghdad who sent his servant to market to buy provisions and in a little while the servant came back, white and trembling, and said, Master, just now when I was in the market-place I was jostled by a woman in the crowd and when I turned I saw it was Death that jostled me. She looked at me and made a threatening gesture; now, lend me your horse, and I will ride away from this city and avoid my fate. I will go to Samarra and there Death will not find me. The merchant lent him his horse, and the servant mounted it, and he dug his spurs in its flanks and as fast as the horse could gallop he went. Then the merchant went down to the market-place and he saw me standing in the crowd, and he came to me and said, why did you make a threatening gesture to my servant when you saw him this morning? That was not a threatening gesture, I said, it was only a start of surprise. I was astonished to see him in Baghdad, for I had an appointment with him tonight in Samarra.

W. Somerset Maugham

In the social sciences, talk about *belief* is ubiquitous. Since social scientists are typically self-conscious about their methods, there is also a lot of talk about *talk about belief*. And since belief is a genuinely curious and perplexing phenomenon, showing many different faces to the world, there is abundant controversy. Sometimes belief attribution appears to be a dark, risky, and imponderable business—especially when exotic, and more particularly religious or superstitious, beliefs are in the limelight. These are not the only troublesome cases; we also court argument and scepticism when we attribute beliefs to non-human animals, or to infants, or to computers or robots. Or when the beliefs we feel constrained to attribute to an apparently healthy, adult member of our own society are contradictory, or even just wildly false. A biologist colleague of mine was once called on the telephone by a man in a bar who wanted him to settle a bet. The man asked: 'Are rabbits birds?' 'No' said the biologist. 'Damn!' said the man as he hung up. Now could he *really* have believed that rabbits were birds? Could anyone really and truly be attributed that belief? Perhaps, but it would take a bit of a story to bring us to accept it.

In all of these cases belief attribution appears beset with subjectivity, infected with cultural relativism, prone to 'indeterminacy of radical translation'—clearly an enterprise demanding special talents: the art of phenomenological analysis, hermeneutics, empathy, *Verstehen*, and all that. On other occasions, normal occasions, when familiar beliefs are the topic, belief attribution looks as easy as speaking prose, and as objective and reliable as counting beans in a dish. Particularly when these straightforward cases are before us, it is quite plausible to suppose that *in principle* (if not yet in practice) it would be possible to confirm these simple, objective belief attributions by *finding something inside the believer's head*—by finding the beliefs themselves, in effect. 'Look', someone might say, 'You either believe there's milk in the fridge or you don't believe there's milk in the fridge' (you might have no opinion, in the latter case). But if you do believe this, that's a perfectly objective fact about you, and it must come down in the end to your brain's being in some particular physical state. If we knew more about physiological psychology we could in principle determine the facts about your brain state, and thereby determine whether or not you believe there is milk in the fridge, even if you were determined to be silent, or disingenuous on the topic. In principle, on this view physiological psychology could trump the results—or non-results—of any 'black box' method in the social sciences that divines beliefs (and other mental features) by behavioural, cultural, social, historical, *external* criteria.

These differing reflections congeal into two opposing views on the nature of belief attribution, and hence on the nature of belief. The latter, a variety of *realism*, likens the question of whether a person has a particular belief to the question of whether a person is infected with a particular virus—a perfectly objective internal matter of fact about which an observer can often make educated guesses of great reliability. The former, which we could call *interpretationism* if we absolutely had to give it a name, likens the question of whether a person has a particular belief to the question of whether a person is immoral, or has style, or talent, or would make a good wife. Faced with such questions, we preface our answers with 'Well, it all depends on what you're interested in', or make some similar acknowledgment of the relativity of the issue. 'It's a matter of interpretation', we say. These two opposing views, so baldly stated, do not fairly represent any serious theorists' positions, but they do express views that are typically seen as mutually exclusive and exhaustive; the theorist must be friendly with one and only one of these themes.

I think this is a mistake. My thesis will be that while belief is a perfectly objective phenomenon (that apparently makes me a realist), it can be discerned only from the point of view of one who adopts a certain *predictive strategy*, and its existence can be confirmed only by an assessment of the success of that strategy (that apparently makes me an interpretationist).

First I will describe the strategy, which I call the intentional strategy, or adopting the intentional stance. To a first approximation, the intentional strategy consists of treating the object whose behaviour you want to predict as a rational agent with beliefs and desires and other mental states exhibiting what Brentano and others call *intentionality*. The strategy has often been described before, but I shall try to put this very familiar material in a new light, by showing *how* it works, and by showing *how well* it works.

Then I will argue that any object—or as I shall say, any *system*—whose behaviour is well predicted by this strategy is in the fullest sense of the word a believer. *What it is* to be a true believer is to be an *intentional system*, a system whose behaviour is reliably and voluminously predictable via the intentional strategy. I have argued for this position before,[1] and my arguments have so far garnered few converts and many presumed counterexamples. I shall try again here, harder, and shall also deal with several compelling objections.

The intentional strategy and how it works

There are many strategies, some good, some bad. Here is a strategy, for instance, for predicting the future behaviour of a person: determine the date and hour of the person's birth, and then feed this modest datum into one or another astrological algorithm for generating predictions of the person's prospects. This strategy is deplorably popular. Its popularity is deplorable only because we have such good reasons for believing that *it does not work.*[2] When astrological predictions come true this is sheer luck, or the result of such vagueness or ambiguity in the prophecy that almost any eventuality can be construed to confirm it. But suppose the astrological strategy did in fact work well on some people. We could call those people *astrological systems*—systems whose behaviour was, as a matter of fact, predictable by the astrological strategy. If there were such people, such astrological systems, we would be more interested than most of us in fact are in *how the astrological strategy works*—that is, we would be interested in the rules, principles, or methods of astrology. We could find out how the strategy works by asking astrologers, reading their books, and observing them in action. But we would also be curious about *why* it worked. We might find that astrologers had no useful opinions about this latter question—they either had no theory of why it worked, or their theories were pure hokum. Having a good strategy is one thing; knowing why it works is another.

So far as we know, however, the class of astrological systems is empty, so the astrological strategy is of interest only as a social curiosity. Other strategies have better credentials. Consider the physical strategy, or physical stance: if you want to predict the behaviour of a system, determine its physical constitution (perhaps all the way down to the micro-physical level) and the physical nature of the impingements upon it, and use your knowledge of the laws of physics to predict the outcome for any input. This is the grand and impractical strategy of Laplace for predicting the entire future of everything in the universe, but it has more modest, local, actually usable versions. The chemist or physicist in the laboratory can use this strategy to predict the behaviour of exotic materials, but equally the cook in the kitchen can predict the effect of leaving the pot on the burner too long. The strategy is not always practically available, but that it will always work *in principle* is a dogma of the physical sciences. (I ignore the minor complications raised by the sub-atomic indeterminacies of quantum physics.)

Sometimes, in any event, it is more effective to switch from the physical stance to what I call the design stance, where one ignores the actual (possibly messy) details of the physical constitution of an object, and, on the assumption that it has a certain design, predicts that it will behave *as it is designed to behave* under various circumstances. For instance, most users of computers have not the foggiest idea what physical principles are responsible for the computer's highly reliable, and hence predictable, behaviour. But if they have a good idea of what the computer is designed to do (a description of its operation at any one of the many possible levels of abstraction), they can predict its behaviour with great accuracy and reliability, subject to disconfirmation only in cases of physical malfunction. Less dramatically, almost anyone can predict when an alarm clock will sound on the basis of the most casual inspection of its exterior. (One does not know or care to know whether it is spring wound, battery driven, sunlight powered, made of brass wheels and jewel bearings or silicon chips—one just assumes that it is designed so that the alarm will sound when it is set to sound, and it is set to sound where it appears to be set to sound, and the clock will keep on running until that time and beyond, and is designed to run more or less accurately, and so forth. For more accurate and detailed design stance predictions of the alarm clock, one must descend to a less abstract level of description of its design; for instance, to the level at which gears are described, but their material is not specified.

Only the designed behaviour of a system is predictable from the design stance, of course. If you want to predict the behaviour of an alarm clock when it is pumped full of liquid helium, revert to the physical stance. Not just artefacts, but also many biological objects (plants and animals, kidneys and hearts, stamens and pistils) behave in ways that can be predicted from the design stance. They are not just physical systems but designed systems.

Sometimes even the design stance is practically inaccessible, and then there is yet another stance or strategy one can adopt: the intentional stance. Here is how it works: first you decide to treat the object whose behaviour is to be predicted as a rational agent; then you figure out what beliefs that agent ought to have, given its place in the world and its purpose. Then you figure out what desires it ought to have, on the same considerations, and finally you predict that this rational agent will act to further its goals in the light of its beliefs. A little practical reasoning from the chosen set of beliefs and desires will in many—but not all—instances yield a decision about what the agent ought to do; that is what you predict the agent *will* do.

The strategy becomes clearer with a little elaboration. Consider first how we go about populating each other's heads with beliefs. A few truisms: sheltered people tend to be ignorant; if you expose someone to something he comes to know all about it. In general, it seems, we come to believe all the truths about the parts of the world around us we are put in a position to learn about. *Exposure* to x, that is, sensory confrontation with x over some suitable period of time, is the *normally sufficient* condition for knowing (or having true beliefs) about x. As we say, we come to *know all about* the things around us. Such exposure is only *normally* sufficient for knowledge, but this is not the large escape hatch it might appear; our threshold for accepting abnormal ignorance in the face of exposure is quite high. 'I didn't know the gun was loaded', said by one who was observed to be present, sighted, and awake during the loading, meets with a variety of utter scepticism that only the most outlandish supporting tale could overwhelm.

Of course we do not come to learn or remember all the truths our sensory histories avail us. In spite of the phrase 'know all about', what we come to know, normally, are only all the *relevant* truths our sensory histories avail us. I do not typically come to know the ratio of spectacle-wearing people to trousered people in a room I inhabit, though if this interested me, it would be readily learnable. It is not just that some facts about my environment are below my thresholds of discrimination or beyond the integration and holding-power of my memory (such as the height in inches of all the people present), but that many perfectly detectable, graspable, memorable facts are of no interest to me, and hence do not come to be believed by me. So one rule for attributing beliefs in the intentional strategy is this: attribute as beliefs all the truths relevant to the system's interests (or desires) that the system's experience to date has made available. This rule leads to attributing somewhat too much–since we all are somewhat forgetful, even of important things. It also fails to capture the false beliefs we are all known to have. But the attribution of false belief, *any* false belief, requires a special genealogy, which will be seen to consist in the main in true beliefs. Two paradigm cases: S believes (falsely) that *p*, because S believes (truly) that Jones told him that *p*, that Jones is pretty clever, that Jones did not intend to deceive him, . . . etc. Second case: S believes (falsely) that there is a snake on the barstool, because S believes (truly) that he seems to see a snake on the barstool, is himself sitting in a bar not a yard from the barstool he sees, and so forth. The falsehood has to start somewhere; the seed may be sown in hallucination, illusion, a normal variety of simple misperception, memory deterioration, or deliberate fraud, for instance, but the false beliefs that are reaped grow in a culture medium of true beliefs.

Then there are the arcane and sophisticated beliefs, true and false, that are so often at the focus of attention in discussions of belief attribution. They do not arise directly, goodness knows, from exposure to mundane things and events, but their attribution requires tracing out a lineage of mainly good argument or reasoning from the bulk of beliefs already attributed. An implication of the intentional strategy, then, is that true believers mainly believe truths. If anyone could devise an agreed-upon method of individuating and counting beliefs (which I doubt very much), we would see that all but the smallest portion (say, less than 10 per cent) of a person's beliefs were attributable under our first rule.[3]

Note that this rule is a derived rule, an elaboration and further specification of the fundamental rule: attribute those beliefs the system *ought to have*. Note also that the rule interacts with the attribution of desires. How do we attribute the desires (preferences, goals, interests) on whose basis we will shape the list of beliefs? We attribute the desires the system *ought to have*. That is the fundamental rule. It dictates, on a first pass, that we attribute the familiar list of highest, or most basic, desires to people: survival, absence of pain, food, comfort, procreation, entertainment. Citing any one of these desires typically terminates the 'Why?' game of reason giving. One is not supposed to need an ulterior motive for desiring comfort or pleasure or the prolongation of one's existence. Derived rules of desire attribution interact with belief attributions. Trivially, we have the rule: attribute desires for those things a system believes to be good for it. Somewhat more informatively, attribute desires for those things a system believes to be best means to other ends it desires. The attribution of bizarre and detrimental desires thus requires, like the attribution of false beliefs, special stories.

The interaction between belief and desire becomes trickier when we consider what desires we attribute on the basis of verbal behaviour. The capacity to *express* desires in language opens the floodgates of desire attribution. 'I want a two-egg mushroom omelette, some French bread and butter, and a half bottle of lightly chilled white Burgundy.' How could one begin to attribute a desire for anything so specific in the absence of such verbal declaration? How, indeed, could a creature come to *contract* such a specific desire without the aid of language? Language *enables* us to formulate highly specific desires, but it also *forces* us on occasion to commit ourselves to desires altogether more stringent in their conditions of satisfaction than anything we would otherwise have any reason to endeavour to satisfy. Since in order to get what you want you often have to say what you want, and since you often cannot say what you want without saying something more specific than you antecedently mean, you often end up giving others evidence—the very best of evidence, your unextorted word—that you desire things or states of affairs far more particular than would satisfy you—or better, than would have satisfied you, for once you have declared, being a man of your word, you acquire an interest in satisfying exactly the desire you declared and no other.

'I'd like some baked beans, please.'

'Yes sir. How many?'

You might well object to having such a specification of desire demanded of you, but in fact we are all socialized to accede to similar requirements in daily life—to the point of not noticing it, and certainly not feeling oppressed by it. I dwell on this because it has a parallel in the realm of belief, where our linguistic environment is forever forcing us to give—or concede—precise verbal expression to convictions that lack the hard edges verbalization endows them with.[4] By concentrating on the *results* of this social force, while ignoring its distorting effect, one can easily be misled into thinking that it is *obvious* that beliefs and desires are rather like *sentences stored in the head*. Being language-using creatures, it is inevitable that we should often come to believe that some particular, actually formulated, spelled and punctuated sentence *is true*, and that on other occasions we should come to want such a sentence to *come true*, but these are special cases of belief and desire, and as such may not be reliable models for the whole domain.

That is enough, on this occasion, about the principles of belief and desire attribution to be found in the intentional strategy. What about the rationality one attributes to an intentional system? One starts with the ideal of perfect rationality and revises downwards as circumstances dictate. That is, one starts with the assumption that people believe all the implications of their beliefs, and believe no contradictory pairs of beliefs. This does not create a practical problem of clutter (infinitely many implications, for instance), for one is interested only in ensuring that the system one is predicting is rational enough to get to the particular implications that are relevant to its behavioural predicament of the moment. Instances of irrationality, or of finitely powerful capacities of inference, raise particularly knotty problems of interpretation, which I will set aside on this occasion.[5]

For I want to turn from the description of the strategy to the question of its use. Do people actually use this strategy? Yes, all the time. There may someday be other strategies for attributing belief and desire and for predicting behaviour, but this is the only one we all know now. And when does it work? It works with people almost all the time. Why would it *not* be a good idea to allow individual Oxford colleges to create and grant academic degrees whenever they saw fit? The answer is a long story, but very easy to generate. And there would be widespread agreement about the major points. We have no difficulty thinking of the reasons people would then have for acting in such ways as to give others reasons for acting in such ways as to give others reasons for . . . creating a circumstance we would not want. Our use of the intentional strategy is so habitual and effortless that the role it plays in shaping our expectations about people is easily overlooked. The strategy also works on most other mammals most of the time. For instance, you can use it to design better traps to catch those mammals, by reasoning about what the creature knows or believes about various things, what it prefers, what it wants to avoid. The strategy works on birds, and on fish, and on reptiles, and on insects and spiders, and even on such lowly and unenterprising creatures as clams (once a clam believes there is danger about, it will not relax its grip on its closed shell until it is convinced that the danger has passed). It also works on some artefacts: the chess-playing computer will not take your knight because it knows that there is a line of ensuing play that would lead to losing its rook, and it does not want that to happen. More modestly, the thermostat will turn off the boiler as soon as it comes to believe the room has reached the desired temperature.

The strategy even works for plants. In a locale with late spring storms you should plant apple varieties that are particularly *cautious* about *concluding* that it is spring—which is when they *want* to blossom, of course. It even works for such inanimate and apparently undesigned phenomena as lightning. An electrician once explained to me how he worked out how to protect my underground water pump from lightning damage: lightning, he said, always wants to find the best way to ground —or earth, as you say in England—but sometimes it gets tricked into taking second-best paths. You can protect the pump by making another, better path more *obvious* to the lightning.

True believers as intentional systems

Now clearly this is a motley assortment of 'serious' belief attributions, dubious belief attributions, pedagogically useful metaphors, *façons de parler*, and perhaps worse: outright frauds. The next task would seem to be distinguishing those intentional systems that *really* have beliefs and desires from those we may find it handy to treat *as if* they had beliefs and desires. But that would be a Sisyphean labour, or else would be terminated by fiat. A better understanding of the phenomenon of belief begins with the observation that even in the worst of these cases, even when we are surest that the strategy works *for the wrong reasons*, it is nevertheless true that it does work, at least a little bit. This is an interesting fact, which distinguishes this class of objects, the class of *intentional systems*, from the class of objects for which the strategy never works. But is this so? Does our definition of an intentional system exclude any objects at all? For instance, it seems the lectern in this lecture room can be construed as an intentional system, fully rational, and believing that it is currently located at the centre of the civilized world (as some of you may also think); and desiring above all else to remain at that centre. What should such a rational agent so equipped with belief and desire do? Stay put, clearly, which is just what the lectern does. I predict the lectern's behaviour, accurately, from the intentional stance, so is it an intentional system? If it is, anything at all is.

What should disqualify the lectern? For one thing, the strategy does not recommend itself in this case, for we get no predictive power from it that we did not antecedently have. We already knew what the lectern was going to do—namely nothing—and tailored the beliefs and desires to fit in a quite unprincipled way. In the case of people, or animals, or computers, however, the situation is different. In these cases often the only strategy that is at all practical is the intentional strategy; it gives us predictive power we can get by no other method. But, it will be urged, this is no difference in nature, but merely a difference that reflects upon our limited capacities as scientists. The Laplacean omniscient physicist could predict the behaviour of a computer—or of a live human body, assuming it to be ultimately governed by the laws of physics—without any need for the risky, short-cut methods of either the design or intentional strategies. For people of limited mechanical aptitude, the intentional interpretation of a simple thermostat is a handy and largely innocuous crutch, but the engineers among us can quite fully grasp its internal operation without the aid of this anthropomorphizing. It may be true that the cleverest engineers find it practically impossible to maintain a clear conception of more complex systems, such as a time-sharing computer system or remote-controlled space probe, without lapsing into an intentional stance (and viewing these devices as asking and telling, trying and avoiding, wanting and believing), but this is just a more advanced case of human epistemic frailty. We would not want to classify these artefacts with the true believers—ourselves—on such variable and parochial grounds, would we? Would it not be intolerable to hold that some artefact, or creature, or person was a believer from the point of view of one observer, but not a believer at all from the point of view of another, cleverer observer? That would be a particularly radical version of interpretationism, and some have thought I espoused it in urging that belief be viewed in terms of the success of the intentional strategy. I must confess that my presentation of the view has sometimes invited that reading, but I now want to discourage it. The decision to adopt the intentional stance is free, but the facts about the success or failure of the stance, were one to adopt it, are perfectly objective.

Once the intentional strategy is in place, it is an extraordinarily powerful tool in prediction—a fact that is largely concealed by our typical concentration on the cases in which it yields dubious or unreliable results. Consider, for instance, predicting moves in a chess game. What makes chess an interesting game, one can see, is the *un*predictability of one's opponent's moves, except in those cases where moves are 'forced'—where there is *clearly* one best move—typically the least of the available evils. But this unpredictability is put in context when one recognizes that in the typical chess situation there are very many perfectly legal and hence available moves, but only a few—perhaps half a dozen—with anything to be said for them, and hence only a few high probability moves according to the intentional strategy. Even where the intentional strategy fails to distinguish a single move with a highest probability, it can dramatically reduce the number of live options.

The same feature is apparent when the intentional strategy is applied to 'real world' cases. It is notoriously unable to predict the exact purchase and sell decisions of stock traders, for instance, or the exact sequence of words a politician will utter when making a scheduled speech, but one's confidence can be very high indeed about slightly less specific predictions: that the particular trader *will not buy utilities today*, or that the politician *will side with the unions against his party*, for example. This inability to predict fine-grained descriptions of actions, looked at another way, is a source of strength for the intentional strategy, for it is this neutrality with regard to details of implementation that permits one to exploit the intentional strategy in complex cases, for instance, in *chaining predictions*.[6] Suppose the US Secretary of State were to announce he was a paid agent of the KGB. What an unparalleled event! How unpredictable its consequences! Yet in fact we can predict dozens of not terribly interesting but perfectly salient consequences, and consequences of consequences. The President would confer with the rest of the Cabinet, which would support his decision to relieve the Secretary of State of his duties pending the results of various investigations, psychiatric and political, and all this would be reported at a news conference to people who would write stories that would be commented upon in editorials that would be read by people who would write letters to the editors, and so forth. None of that is daring prognostication, but note that it describes an arc of causation in space-time that could not be predicted under *any* description by any imaginable practical extension of physics or biology.

The power of the intentional strategy can be seen even more sharply with the aid of an objection first raised by Robert Nozick some years ago. Suppose, he suggested, some beings of vastly superior intelligence—from Mars, let us say—were to descend upon us, and suppose that we were to them as simple thermostats are to clever engineers. Suppose, that is, that they did not *need* the intentional stance—or even the design stance—to predict our behaviour in all its detail. They can be supposed to be Laplacean super-physicists, capable of comprehending the activity on Wall Street, for instance, at the micro-physical level. Where we see brokers and buildings and sell orders and bids, they see vast congeries of sub-atomic particles milling about—and they are such good physicists that they can predict days in advance what ink marks will appear each day on the paper tape labelled 'Closing Dow Jones Industrial Average'. They can predict the individual behaviours of all the various moving bodies they observe without ever treating any of them as intentional systems. Would we be right then to say that from *their* point of view we really were not believers at all (any more than a simple thermostat is)? If so, then our status as believers is nothing objective, but rather something in the eye of the beholder—provided the beholder shares our intellectual limitations.

Our imagined Martians might be able to predict the future of the human race by Laplacean methods, but if they did not also see us as intentional systems, they would be *missing something* perfectly objective: the *patterns* in human behaviour that are describable from the intentional stance, and only from that stance, and which support generalizations and predictions. Take a particular instance in which the Martians observe a stock broker deciding to place an order for 500 shares of General Motors. They predict the exact motions of his fingers as he dials the phone, and the exact vibrations of his vocal cords as he intones his order. But if the Martians do not see that indefinitely many *different* patterns of finger motions and vocal cord vibrations—even the motions of indefinitely many different individuals—could have been substituted for the actual particulars without perturbing the subsequent operation of the market, then they have failed to see a real pattern in the world they are observing. Just as there are indefinitely many ways of *being a spark plug*—and one has not understood what an internal combustion engine is unless one realizes that a variety of different devices can be screwed into these sockets without affecting the performance of the engine—so there are indefinitely many ways of *ordering 500 shares of General Motors*, and there are societal sockets in which one of these ways will produce just about the same effect as any other. There are also societal pivot points, as it were, where which way people go depends on whether they *believe that p*, or *desire A*, and does not depend on any of the other infinitely many ways they may be alike or different.

Suppose, pursuing our Martian fantasy a little further, that one of the Martians were to engage in a predicting contest with an Earthling. The Earthling and the Martian observe (and observe each other observing) a particular bit of local physical transaction. From the Earthling's point of view, this is what is observed. The telephone rings in Mrs Gardner's kitchen. She answers, and this is what she says: 'Oh, hello dear. You're coming home early? Within the hour? And bringing the boss to dinner? Pick up a bottle of wine on the way home, then, and drive carefully.' On the basis of this observation, our Earthling predicts that a large metallic vehicle with rubber tyres will come to a stop in the drive within one hour, disgorging two human beings one of whom will be holding a paper bag containing a bottle containing an alcoholic fluid. The prediction is a bit risky, perhaps, but a good bet on all counts. The Martian makes the same prediction, but has to avail himself of much more information about an extraordinary number of interactions of which, so far as he can tell, the Earthling is entirely ignorant. For instance, the deceleration of the vehicle at intersection *A*, five miles from the house, without which there would have been a collision with another vehicle—whose collision course had been labouriously calculated over some hundreds of metres by the Martian. The Earthling's performance would look like magic! How did the Earthling know that the human being who got out of the car and got the bottle in the shop would get back in? The coming true of the Earthling's prediction, after all the vagaries, intersections, and branches in the paths charted by the Martian, would seem to anyone bereft of the intentional strategy as marvellous and inexplicable as the fatalistic inevitability of the appointment in Samarra. Fatalists—for instance, astrologers—believe that there is a pattern in human affairs that is inexorable, that will impose itself *come what may*, that is, no matter how the victims scheme and second-guess, no matter how they twist and turn in their chains. These fatalists are wrong, but they are *almost* right. There *are* patterns in human affairs that impose themselves, not quite inexorably but with great vigour, absorbing physical perturbations and variations that might as well be considered random; these are the patterns that we characterize in terms of the beliefs, desires, and intentions of rational agents.

No doubt you will have noticed, and been distracted by, a serious flaw in our thought experiment: the Martian is presumed to treat his Earthling opponent as an intelligent being like himself, with whom communication is possible, a being with whom one can make a wager, against whom one can compete. In short, a being with beliefs (such as the belief he expressed in his prediction) and desires (such as the desire to win the prediction contest). So if the Martian sees the pattern in one Earthling, how can he fail to see it in the others? As a bit of narrative, our example could be strengthened by supposing that our Earthling cleverly learned Martian (which is transmitted by X-ray modulation) and disguised himself as a Martian, counting on the species-chauvinism of these otherwise brilliant aliens to permit him to pass as an intentional system while not giving away the secret of his fellow human beings. This addition might get us over a bad twist in the tale, but might obscure the moral to be drawn: namely, the unavoidability of the intentional stance with regard to oneself and one's fellow intelligent beings. This unavoidability is itself interest relative; it is perfectly possible to adopt a physical stance, for instance, with regard to an intelligent being, oneself included, but not to the exclusion of maintaining at the same time an intentional stance with regard to oneself at a minimum, and one's fellows *if* one intends, for instance, to learn what they know (a point that has been powerfully made by Stuart Hampshire in a number of writings). We can perhaps suppose our superintelligent Martians fail to recognize *us* as intentional systems, but we cannot suppose them to lack the requisite concepts.[7] If they observe, theorize, predict, communicate, they view *themselves* as intentional systems.[8] Where there are intelligent beings the patterns must be there to be described, whether or not we care to see them.

It is important to recognize the objective reality of the intentional patterns discernible in the activities of intelligent creatures, but also important to recognize the incompleteness and imperfections in the patterns. The objective fact is that the intentional strategy *works as well as it does*, which is not perfectly. No one is perfectly rational, perfectly unforgetful, all-observant, or invulnerable to fatigue, malfunction, or design imperfection. This leads inevitably to circumstances beyond the power of the intentional strategy to describe, in much the same way that physical damage to an artefact, such as a telephone or an automobile, may render it indescribable by the normal design terminology for that artefact. How do you draw the schematic wiring diagram of an audio amplifier that has been partially melted, or how do you characterize the programme state of a malfunctioning computer? In cases of even the mildest and most familiar cognitive pathology—where people seem to hold contradictory beliefs, or to be deceiving themselves, for instance—the canons of interpretation of the intentional strategy fail to yield clear, stable verdicts about which beliefs and desires to attribute to a person.

Now a *strong* realist position on beliefs and desires would claim that in these cases the person in question really does have some particular beliefs and desires which the intentional strategy, as I have described it, is simply unable to divine. On the milder sort of realism I am advocating, there is no fact of the matter of exactly which beliefs and desires a person has in these degenerate cases, but this is not a surrender to relativism or subjectivism, for *when* and *why* there is no fact of the matter is itself a matter of objective fact. On this view one can even acknowledge the *interest relativity* of belief attributions, and grant that given the different interests of different cultures, for instance, the beliefs and desires one culture would attribute to a member might be quite different from the beliefs and desires another culture would attribute to that very same person. But supposing that were so in a particular case, there would be the further facts about *how well* each of the rival intentional strategies worked for predicting the behaviour of that person. We can be sure in advance that no intentional interpretation of an individual will work to perfection, and it may be that two rival schemes are about equally good, and better than any others we can devise. That this is the case is itself something about which there can be a fact of the matter. The objective presence of one pattern (with whatever imperfections) does not rule out the objective presence of another pattern (with whatever imperfections).

The bogey of radically different interpretations with equal warrant from the intentional strategy is theoretically important—one might better say metaphysically important–but practically negligible once one restricts one's attention to the largest and most complex intentional systems we know: human beings.[9]

Until now I have been stressing our kinship to clams and thermostats, in order to emphasize a view of the logical status of belief attribution, but the time has come to acknowledge the obvious differences, and say what can be made of them. The perverse claim remains: *all there is* to being a true believer is being a system whose behaviour is reliably predictable via the intentional strategy, and hence *all there is* to really and truly believing that p (for any proposition p) is being an intentional system for which p occurs as a belief in the best (most predictive) interpretation. But once we turn our attention to the truly interesting and versatile intentional systems, we see that this apparently shallow and instrumentalistic criterion of belief puts a severe constraint on the internal constitution of a genuine believer, and thus yields a robust version of belief after all.

Consider the lowly thermostat, as degenerate a case of an intentional system as could conceivably hold our attention for more than a moment. Going along with the gag we might agree to grant it the capacity for about half a dozen different beliefs and fewer desires–it can believe the room is too cold or too hot, that the boiler is on or off, and that if it wants the room warmer it should turn on the boiler, and so forth. But surely this is imputing too much to the thermostat; it has no concept of heat or of a boiler, for instance. So suppose we *de-interpret* its beliefs and desires: it can believe the A is too F or G, and if it wants the A to be more F it should do K, and so forth. After all, by attaching the thermostatic control mechanism to different input and output devices, it could be made to regulate the amount of water in a tank, or the speed of a train, for instance. Its attachment to a heat-sensitive 'transducer' and a boiler is too impoverished a link to the world to grant any rich semantics to its belief-like states.

But suppose we then enrich these modes of attachment. Suppose we give it more than one way of learning about the temperature, for instance. We give it an eye of sorts that can distinguish huddled, shivering occupants of the room, and an ear so that it can be told how cold it is. We give it some facts about geography so that it can conclude that it is probably in a cold place if it learns that its spatio-temporal location is Winnipeg in December. Of course giving it a visual system that is multi-purpose and general–not a mere shivering-object detector–will require vast complications of its inner structure. Suppose we also give our system more behavioural versatility: it chooses the boiler fuel, purchases it from the cheapest and most reliable dealer, checks the weather stripping and so forth. This adds another dimension of internal complexity; it gives individual belief-like states *more to do*, in effect, by providing more and different occasions for their derivation or deduction from other states, and by providing more and different occasions for them to serve as premises for further reasoning. The cumulative effect of enriching these connections between the device and the world in which it resides is to enrich the semantics of its dummy predicates, F and G and the rest. The more of this we add, the less amenable our device becomes to serving as the control structure of anything other than a room temperature maintenance system. A more formal way of saying this is that the class of indistinguishably satisfactory models of the formal system embodied in its internal states gets smaller and smaller as we add such complexities; the more we add, the richer or more demanding or specific the semantics of the system, until eventually we reach systems for which a *unique* semantic interpretation is *practically* (but never *in principle*) dictated.[10] At that point we say this device (or animal, or person) has beliefs *about heat*, and *about this very room*, and so forth, not only because of the system's *actual* location in, and operations on, the world, but because we cannot imagine another niche in which it could be placed *where it would work*.

Our original simple thermostat had a state we called a belief about a particular boiler, to the effect that it was on or off. Why about *that* boiler? Well, what *other* boiler would you want to say it was about? The belief is about the boiler because it is *fastened* to the boiler.[11] Given the actual, if minimal, causal link to the world that happened to be in effect, we could endow a state of the device with *meaning* (of a sort) and *truth conditions*, but it was altogether too easy to substitute a different minimal link and completely change the meaning (in this impoverished sense) of that internal state. But as systems become perceptually richer and behaviourally more versatile, it becomes harder and harder to make substitutions in the actual links of the system to the world without changing the organization of the system itself. If you change its environment, it will *notice*, in effect, and make a change in its internal state in response. There comes to be a two-way constraint of growing specificity between the device and the environment. Fix the device in any one state and it demands a *very* specific environment in which to operate properly (you can no longer switch it easily from regulating temperature to regulating speed or anything else); but at the same time, if you do not *fix* the state it is in, but just plonk it down in a changed environment, its sensory attachments will be sensitive and discriminative enough to respond appropriately to the change, driving the system into a new state, in which it will operate effectively in the new environment. There is a familiar way of alluding to this tight relationship that can exist between the organization of a system and its environment: you say that the organism continuously *mirrors* the environment, or that there is a *representation* of the environment in—or implicit in—the organization of the system.

It is not that we attribute (or should attribute) beliefs and desires only to things in which we find internal representations, but rather that when we discover some object for which the intentional strategy works, we endeavour to interpret some of its internal states or processes as internal representations. What makes some internal feature of a thing a representation could only be its role in regulating the behaviour of an intentional system.

Now the reason for stressing our kinship with the thermostat should be clear. There is no magic moment in the transition from a single thermostat to a system that *really* has an internal representation of the world around it. The thermostat has a minimally demanding representation of the world, fancier thermostats have more demanding representations of the world, fancier robots for helping around the house would have still more demanding representations of the world. Finally you reach us. We are so multifariously and intricately connected to the world that almost no substitution is possible—though it is clearly imaginable in a thought experiment. Hilary Putnam imagines the planet Twin Earth, which is just like Earth right down to the scuff marks on the shoes of the Twin Earth replica of your neighbour, but which differs from Earth in some property that is entirely beneath the thresholds of your capacities to discriminate. (What they call water on Twin Earth has a different chemical analysis.) Were *you* to be whisked instantaneously to Twin Earth and exchanged for your Twin Earth replica, you would never be the wiser—just like the simple control system that cannot tell whether it is regulating temperature, speed, or volume of water in a tank. It is easy to devise radically different Twin Earths for something as simple and sensorily deprived as a thermostat, but your internal organization puts a much more stringent demand on substitution. Your Twin Earth and Earth must be virtual replicas or you will change state dramatically on arrival.

So which boiler are *your* beliefs about, when you believe the boiler is on? Why, the boiler in your cellar (rather than its twin on Twin Earth, for instance). What *other* boiler would your beliefs be about? The *completion* of the semantic interpretation of your beliefs, fixing the *referents* of your beliefs, requires, as in the case of the thermostat, facts about your actual embedding in the world. The principles, and problems, of interpretation that we discover when we attribute beliefs to people are the *same* principles and problems we discover when we look at the ludicrous, but blessedly simple, problem of attributing beliefs to a thermostat. The differences are of degree, but nevertheless of such great degree that understanding the internal organization of a simple intentional system gives one very little basis for understanding the internal organization of a complex intentional system, such as a human being.

Why does the intentional strategy work?

When we turn to the question of *why* the intentional strategy works as well as it does, we find that the question is ambiguous, admitting of two very different sorts of answers. If the intentional system is a simple thermostat, one answer is simply this: the intentional strategy works because the thermostat is well designed; it was designed to be a system that could be easily and reliably comprehended and manipulated from this stance. That is true, but not very informative, if what we are after are the actual features of its design that explain its performance. Fortunately, however, in the case of a simple thermostat those features are easily discovered and understood, so the other answer to our *why* question, which is really an answer about *how the machinery works*, is readily available.

If the intentional system in question is a person, there is also an ambiguity in our question. The first answer to the question of why the intentional strategy works is that evolution has designed human beings to be rational, to believe what they ought to believe and want what they ought to want. The fact that we are products of a long and demanding evolutionary process guarantees that using the intentional strategy on us is a safe bet. This answer has the virtues of truth and brevity, and on this occasion the additional virtue of being an answer Herbert Spencer would applaud, but it is also strikingly uninformative. The more difficult version of the question asks, in effect, how the machinery which Nature has provided us works. And we cannot yet give a good answer to that question. We just do not know. We do know how the *strategy* works, and we know the easy answer to the question of why it works, but knowing these does not help us much with the hard answer.

It is not that there is any dearth of doctrine, however. A Skinnerian behaviourist, for instance, would say that the strategy works because its imputations of beliefs and desires are shorthand, in effect, for as yet unimaginably complex descriptions of the effects of prior histories of response and reinforcement. To say that someone wants some ice cream is to say that in the past the ingestion of ice cream has been reinforced in him by the results, creating a propensity under certain background conditions (also too complex to describe) to engage in ice-cream-acquiring behaviour. In the absence of detailed knowledge of those historical facts we can nevertheless make shrewd guesses on inductive grounds; these guesses are embodied in our intentional stance claims. Even if all this were true, it would tell us very little about the way such propensities were regulated by the internal machinery.

A currently more popular explanation is that the account of how the strategy works and the account of how the mechanism works will (roughly) *coincide*: for each predictively attributable belief, there will be a functionally salient internal state of the machinery, decomposable into functional parts in just about the same way the sentence expressing the belief is decomposable into parts—that is, words or terms. The inferences we attribute to rational creatures will be mirrored by physical, causal processes in the hardware; the *logical* form of the propositions believed will be copied in the *structural* form of the states in correspondence with them. This is the hypothesis that there is a *language of thought* coded in our brains, and our brains will eventually be understood as symbol manipulating systems in at least rough analogy with computers. Many different versions of this view are currently being explored, in the new research programme called cognitive science, and provided one allows great latitude for attenuation of the basic, bold claim, I think some version of it will prove correct.

But I do not believe that this is *obvious*. Those who think that it is obvious, or inevitable, that such a theory will prove true (and there are many who do), are confusing two different empirical claims. The first is that intentional stance description yields an objective, real pattern in the world—the pattern our imaginary Martians missed. This is an empirical claim, but one that is confirmed beyond scepticism. The second is that this real pattern is *produced by* another real pattern roughly isomorphic to it within the brains of intelligent creatures.

Doubting the existence of the second real pattern is not doubting the existence of the first. There *are* reasons for believing in the second pattern, but they are not overwhelming. The best simple account I can give of the reasons is as follows.

As we ascend the scale of complexity from simple thermostat, through sophisticated robot, to human being, we discover that our efforts to design systems with the requisite behaviour increasingly run foul of the problem of *combinatorial explosion*. Increasing some parameter by, say, 10 per cent—10 per cent more inputs, or more degrees of freedom in the behaviour to be controlled, or more words to be recognized, or whatever—tends to increase the internal complexity of the system being designed by orders of magnitude. Things get out of hand very fast and, for instance, can lead to computer programs that will swamp the largest, fastest machines. Now somehow the brain has solved the problem of combinatorial explosion. It is a gigantic network of billions of cells, but still finite, compact, reliable, and swift, and capable of learning new behaviours, vocabularies, theories, almost without limit. Some elegant, *generative*, indefinitely extendable principles of representation must be responsible. We have only one model of such a representation system: a human language. So the argument for a language of thought comes down to this: what else could it be? We have so far been unable to imagine any plausible alternative in any detail. That is a good enough reason, I think, for recommending as a matter of scientific tactics that we pursue the hypothesis in its various forms as far as we can.[12] But we will engage in that exploration more circumspectly, and fruitfully, if we bear in mind that its inevitable rightness is far from assured. One does not well understand even a true empirical hypothesis so long as one is under the misapprehension that it is necessarily true.[13]

Notes

1. Intentional systems. *Journal of Philosophy* (1971). Conditions of personhood. In *The identities of persons* (ed. A. Rorty). University of California Press (1975). Both reprinted in *Brainstorms* Montgomery, Vt., Bradford (1978). Three kinds of intentional psychology. In *Mind, psychology and reductionism* (ed. R. A. Healey). Cambridge University Press (1981).

2. *Pace* Paul Feyerabend, whose latest book, *Science in a free society*, New Left Books, London (1978), is heroically open-minded about astrology.

3. The idea that most of anyone's beliefs *must* be true seems obvious

to some people. Support for the idea can be found in works by Quine, Putnam, Shoemaker, Davidson, and myself. Other people find the idea equally incredible—so probably each side is calling a different phenomenon belief. Once one makes the distinction between belief and opinion (in my technical sense—see How to change your mind. In *Brainstorms* Chapter 16), according to which opinions are linguistically infected, relatively sophisticated cognitive states—*roughly*, states of betting on the truth of a particular, formulated sentence, one can see the near triviality of the claim that most beliefs are true. A few reflections on peripheral matters should bring it out. Consider Democritus, who had a systematic, all-embracing, but (let us say, for the sake of argument) entirely false physics. He had things *all wrong*, though his views held together and had a sort of systematic utility. But even if every *claim* that scholarship permits us to attribute to Democritus (either explicit or implicit in his writings) is false, these represent a vanishingly small fraction of his *beliefs*, which include both the vast numbers of humdrum standing beliefs he must have had (about which house he lived in, what to look for in a good pair of sandals, and so forth), and also those occasional beliefs that came and went by the millions as his perceptual experience changed.

But, it may be urged, this isolation of his humdrum beliefs from his science relies on an insupportable distinction between truths of observation and truths of theory; all Democritus' beliefs are theory-laden, and since his theory is false, they are false. The reply is as follows: Granted that all observation beliefs are theory laden, why should we choose Democritus' *explicit*, sophisticated theory (couched in his *opinions*) as the theory with which to burden his quotidian observations? Note that the least theoretical compatriot of Democritus also had myriads of theory-laden observation beliefs—and was, in one sense, none the wiser for it. Why should we not suppose their observations are laden with the same theory? If Democritus forgot his theory, or changed his mind, his observational beliefs would be *largely* untouched. To the extent that his sophisticated theory played a discernible role in his routine behaviour and expectations and so forth, it would be quite appropriate to couch his humdrum beliefs in terms of the sophisticated theory, but this will not yield a *mainly false* catalogue of beliefs, since so few of his beliefs will be affected. (The effect of theory on observation is nevertheless often underrated. See Paul Churchland, *Scientific realism and the plasticity of mind*, Cambridge University Press (1979), for dramatic and convincing examples of the tight relationship that can sometimes exist between theory and experience. (The discussion in this note was distilled from a useful conversation with Paul and Patricia Churchland and Michael Stack.)

4. See my *Content and consciousness* pp. 184–5, Routledge & Kegan Paul, London (1969), and How to change your mind, in *Brainstorms*.

5. See *Brainstorms,* and Three kinds of intentional psychology. See also C. Cherniak, Minimal rationality *Mind*, (1981) and my response to Stephen Stich's Headaches. In *Philosophical books*, (1980).

6. See On giving libertarians what they say they want, in *Brainstorms*.

7. A member of the audience in Oxford pointed out that if the Martian included the Earthling in his physical stance purview (a possibility I had not explicitly excluded), he would not be surprised by the Earthling's prediction. He would indeed have predicted exactly the pattern of X-ray modulations produced by the Earthling speaking Martian. True, but as the Martian wrote down the results of his calculations, his prediction of the Earthling's prediction would appear, word by Martian word, as on a Ouija board, and what would be baffling to the Martian was how this chunk of mechanism, the Earthling predictor dressed up like a Martian, was able to yield this *true* sentence of Martian when it was so informationally isolated from the events the Martian needed to know of in order to make his own prediction about the arriving automobile.

8. Might there not be intelligent beings who had no use for communicating, predicting, observing. . . ? There might be marvellous, nifty, invulnerable entities lacking these modes of action, but I cannot see what would lead us to call them *intelligent*.

9. John McCarthy's analogy to cryptography nicely makes this point. The larger the corpus of cipher text, the less chance there is of dual, systematically unrelated decipherings. For a very useful discussion of the principles and presuppositions of the intentional stance applied to machines—explicitly including thermostats—see McCarthy's Ascribing mental qualities to machines. In *Philosophical perspectives on artificial intelligence* (ed. Martin Ringle). Humanities Press (1979).

10. Patrick Hayes explores this application of Tarskian model theory to the semantics of mental representation in The naive physics manifesto (forthcoming).

11. This idea is the ancestor in effect of the species of different ideas lumped together under the rubric of *de re* belief. If one builds from this idea towards its scions one can see better the difficulties with them, and how to repair them.

12. The fact that all *language of thought* models of mental representation so far proposed fall victim to combinatorial explosion in one way or another should temper one's enthusiasm for engaging in what Fodor aptly calls 'the only game in town'.

13. This paper was written during a Fellowship at the Center for Advanced Study in the Behavioral Sciences. I am grateful for financial support provided by the National Endowment for the Humanities, the National Science Foundation (BNS 78-24671), and the Alfred P. Sloan Foundation.

5 *Non-corporeal explanation in psychology*

DONALD E. BROADBENT

Oxford University

I hope that those of you who have some sensitivity to the language were sent leaping to your dictionaries when you saw in the title of this paper the word 'non-corporeal'. If you did, it is almost certain that you were unsuccessful. I have had a look at most of the standard dictionaries, and they give no such usage. They mostly give two other words, 'corporeal' and 'incorporeal'. To take the *Concise Oxford Dictionary* as an example, the entry for 'corporeal' reads 'bodily; material'. The entry for 'incorporeal' reads 'not composed of matter; of immaterial beings'.

Now, in the study of psychology it is trite to point out that explanations are frequently of two types. One kind appeals to physical events or structures; this class of explanation is of growing importance for anybody who cares about human nature. When for example Professor Weiskrantz examines a patient who reports that he has no visual experience, it is a satisfactory explanation of this fact that there is some interruption of the pathways from the eye through the lateral geniculate to the striate cortex. When furthermore the same patient shows that he is capable of reacting to light to some extent, despite the absence of any experience of vision, this in turn is explained by the continued existence of other neural pathways which take information from the eye to other parts of the brain. Similarly, when some of the symptoms of schizophrenia are relieved by treatment with chlorpromazine or other phenothiazines, and particularly when the success of the drug appears to be related to its power to inhibit the action of dopamine, then it would be very strange not to explain the change in behaviour and experience by the physical action of the drug.

In contrast to these corporeal explanations, there are others which are given in psychology, and which do not refer to any physical event or structure. I shall be giving some examples from actual research later on, but for the moment let us simply agree that the explanation of some psychological problem is often to be found in reported experience, in states of feeling, or in declared intentions and purposes. As you all know, this latter class of explanations has frequently been seen as incorporeal, in the sense given by the dictionaries. That is, they are seen by some people as totally divorced from any physical realization, and as belonging to a quite different realm of study. Indeed in previous years it has not been unknown for a Herbert Spencer lecturer to complain of the positivistic bias of academic psychology, since the tendency to demand public evidence in support of an explanation can be seen as hostile to this second class of explanation, the class which cannot be called corporeal.

There is in fact a long-standing division in our culture, between those who favour the first type of explanation and those who favour the second. I want to take two statements as representative of these points of view; as it happens, both statements were made in print by Fellows of the Royal Society, but, if you will forgive me, I will not name the author in either case. The reason is that both statements were made a few years ago, that each of them is taken out of context, may have been written in an atypical mood, and in general that these statements may not represent the considered stance of either author. You will however recognize the attitude and flavour, when I quote one of these distinguished scientists as saying, in a discussion of memory, 'To study the material organisation behind any subject or problem is surely the basis of a scientific approach'. A little later he quotes with approval the supposition of Hooke that ideas are really 'corporeal and material'. In another place he quotes a psychologist who points out that one can acquire information about learning without studying the nervous system, and remarks 'The Royal Society has fortunately not been content to accept such truisms as the last word'. We have here in other words the attitude that corporeal explanation is superior, fundamental, and more scientific.

At the opposite extreme, my other distinguished scientist says 'Neither neurophysiology nor behaviourist psychology will suffice for the construction of a science of the mind, because their concepts are not mental but physical. The initiative must come from a more abstract level of description, such as logic or linguistics'. Here, you see, the appeal is to concepts which have no material connections at all: logic and linguistics are incorporeal.

Some of the developments in the philosophy of science make the appeal to the immaterial seem quite reasonable. After all, somebody might say 'I have a theory that people are logical, and I have read Popper, and I will believe that theory until it is disproved'. When he is then presented with a counter-instance, he might say 'Ah, but I have read Kuhn, and I am working in a different paradigm where your methods are disallowed'. I hasten to say that the Fellow of the Royal Society I am quoting is much too intelligent to argue like that; but he *has* said, elsewhere 'The main problem in linguistics as in other sciences is not so much that of choosing between alternative, equally plausible, theories, but that of constructing any theory at all which will harmonize with the vast range of observable phenomena'.

You will have noticed, in the statements of each point of view, the criticism of professional psychologists, which appears to be the one point on which these two views coincide. That is entirely appropriate, because of course the tradition of my subject is to deny both these positions.

Briefly, then, I want to explain the position that scientifically interesting explanations for psychological phenomena may indeed often be corporeal, but equally may often be of another kind. If one were to confine one's attention to the first kind, the science would be impoverished. When however one uses an explanation of the second kind, this does not mean that one is talking of events detached from any physical entity; on the contrary, explanations of this kind are clearly necessary in dealing with complex physical systems, even ones which are manifestly far simpler than human beings. It is therefore misleading and poor semantic hygiene to refer to explanations without physical content as being 'incorporeal'.

At last therefore you see why I want to use the word 'non-corporeal'. I want to deny the position of those who restrict themselves solely to physiological or biochemical study, because we do need non-corporeal concepts. Conversely, I also want to assert that any actual implementation of a non-corporeal entity requires some physical manifestation, by reference to which we get evidence about the truth or falsehood of the non-corporeal explanation. The validity of some explanation of human reasoning, or our use of language, has to be shown by analysis of the actual success or failure of real people in solving particular problems or understanding particular sentences.

In the case of logic, I do assure you that many careful and solemn experiments have shown that people's minds do not conform to logical principles (in case you had not noticed). In the case of linguistics, there was some years ago the view that human perception of sentences might perhaps exploit grammatical information as represented in the grammars proposed by linguists. This attempt is now broadly agreed to have failed, the most notable test being that those who first supported it (such as Jerry Fodor or George Miller) have very properly accepted the conclusion in print and gone on to new enterprises. There is perhaps a certain return amongst psychologists to the values of positivism, evidence, and the empirical approach. But this does not mean confining ourselves to material explanations.

It is much easier to see what somebody means when they point to concrete illustrations. I want therefore to discuss non-corporeal explanation in the context of a particular real situation. To bring out my claim that such explanation is necessary for systems other than human beings, I am going to do it in the case of computer programs. A little over a year ago, my colleague Peter FitzGerald and I were about to undertake some experiments on the efficiency with which human beings could control simple economic systems. To do this, we represent the economic system by a suitably programmed computer, and ask the human beings to decide on a suitable tax rate and level of government expenditure for the country they are supposed to be governing. When they have made their decisions, the computer calculates the level of unemployment and of inflation which would result from the decision, and announces these results to the decision-maker. The person we are studying then has to make another decision, the results come out again, and so on. The findings of the experiment are of no particular concern here, although I cannot resist saying that we obtained very high inflation rates.

My main point in mentioning the study here, however, is that Peter FitzGerald and I were a little worried in case there was some mistake in the computer program. It was fairly complicated, and even a minor slip such as a typing error might result in the wrong numbers coming out of the machine in response to the subject's decisions. We therefore thought it advisable for each of us separately and independently to write a program which should give the same answer. These programs were, when written, exceedingly different; Peter is a skilful and elegant programmer, whereas I am a relatively incompetent, slipshod, and amateur one. To the knowledgeable, any program written by me creates the same emotional impression as might some attempt by a contemporary sociologist to write a new version of Thomas Jefferson's

Declaration of Independence. If that sad event were to take place, it is likely that the reader would be offended by long and ungainly sentences, by infelicities of phrase, ambiguities of meaning which require very close attention to decipher, and so on. Any program written by me has these characteristics abundantly, and at no point whatever did my program resemble the one written by Peter FitzGerald. The sequence within the computer which occurred when my program was inserted was totally different in every way from that which occurred when Peter's program was inserted. Information of different types was put in different places in the machine, the sequence of operations performed on it was different, and even the time taken by the whole process was different. Yet we confidently hoped that the two programs would give the same numbers in response to any particular decision taken by our experimental subjects, and indeed in the end we did find that they did so. The identity of the physical events was not therefore essential for the identity of the final output. In fact there was a discrepancy at one point, because when we first tried the programs, we found that they did not give the same results, and for this we required an explanation. What kind of explanation would we require?

In contrast to Professor Weiskrantz's patients, the computer had no defects whatever in any of its wiring or circuitry. The physical structure of the machine was perfectly satisfactory. Equally, one could not say that the sequence of physical events within those circuits was the same for both programs up to a particular point where it diverged, because neither sequence had any resemblance to the other, at the level of concrete physical events. In fact, I can go even further than that, because both these programs were written in a language (BASIC) which can be interpreted by a number of different computers. Usually there are minor differences between the particular dialects of a language spoken by machines at different places, but it so happens that the makers of our own machine have recently introduced a different model, which has its memory laid out in a rather different way. Yet it is perfectly capable of accepting the identical programs which can be run on the model we were using. Consequently it would be possible to take my program and put it into the two kinds of machine, and if so the physical events which would occur in the two machines would in detail be very different. When I asked the machine to take in the subject's decision about tax rate, in one machine this information would be put in one part of memory, and in the other machine in a quite different part of memory. The physical events occurring when one implements my program on one machine, would be different from the events which occur when one implements it on another machine. However, in this case the two answers that came out of the machine would be the same. Equally, if Peter's program had been run on the two machines, it would have given the same results with either machine. The discrepancy only arises when one compares the results of one program with the results of the other program.

Not to keep you in suspense, the explanation for the discrepancy was a failure of communication between myself and Peter. When one is constructing a model of an economy, the level of government expenditure is of course stated in terms of the prices ruling at a certain point in time. Between one decision and another, inflation occurs, and the relationship between the amount of government expenditure and the real capacity of the economy changes. In agreeing on the operations the machine has to perform, we had not come to a formal agreement whether we were to regard the demand placed by government expenditure on the economy as stated in the prices holding at the beginning of the decision period, or the prices holding at the end. One of us had done one of these, and the other had done the opposite. As soon as we spotted this point, and both wrote our programs using the same price level, the two programs gave identical results.

My example has resemblances to, but also differences from, the familiar distinction between explanations appropriate to science at different levels. As was said by my second Fellow of the Royal Society, the explanation of the discrepancy we have been considering lies at a very high level of abstraction. It is a familiar idea that one can talk about events in the physical world at different levels. In the nervous system for example one can discuss the passage of an impulse across the synaptic junction between one nerve cell and another. Alternatively, one can look into the chemical processes involved, and see how the transmitter substance is broken down by the appropriate enzyme. Looking more closely still, we can consider the bonds between the various atoms, which lock them into one molecule or another, and the way in which electrons shift their allegiance. Talking about the process at a lower level is of a kind of expansion and translation, which adds a great deal of detail, rather as if one were looking at a crowd seen on television, and then the camera zoomed in to show, first, a discussion going on between a group in the crowd, and then the face and gestures of one particular individual. Concentrating on the individual does not affect his status as a member of the group, or that of the group as a member of the crowd, but is merely convenient for different purposes. Life is too short to detail the behaviour of individual electrons when we are talking about transmission of nervous impulses, so we may as well stick to the short label appropriate to a relatively high-level science. Chemists continue to find plenty of interest in their lives, even though particle physicists may feel that they ought really to be considering more basic matters. It is also true that the discussion at a higher level has to involve some terms, such as statements of relations, which are not used at a lower level; in discussing whether one molecule will react with another, it becomes extremely important whether the spatial relations between the atoms in one substance are so arranged that they are appropriate to the spatial relations between the atoms in the other molecule. (For instance, this is important for the possible role of the phenothiazines in controlling schizophrenia, noted earlier.) For these reasons, most people are happy to accept that sciences of different levels have their own degree of autonomy, and that nevertheless there is a unity of explanation between the whole of science.

I would like to suggest however that the problems of explanation which arise when one considers the discrepancy between two computer programs are interestingly different from those which arise in the familiar case of sciences at different levels. The discrepancy between the two programs exists at the abstract level; but that abstract level can be implemented by many different lower levels. In the usual separation of sciences into levels, we may discuss the properties of water, or we may discuss the ways in which hydrogen and oxygen atoms can link together to form an integrated molecule. Yet a molecule of normal water does consist of two atoms of hydrogen and one of oxygen, and the rules governing implementation of the higher level on the lower are constant. When we deal with descriptions of computer programs, the rules governing implementation are not constant, and the higher level does not simply add a few relational statements to the laws governing the lower level. The implementation may be totally different physically between two programs giving the same result, as well as between two that give a different result. Explanation holds good only at the abstract level.

Even if one grants that this is so, does a similar kind of problem appear in psychology? Certainly it does: if a person suffers serious injury to one side of the brain, as an adult, they lose particular functions which are different depending upon the side that has been damaged. If however the damage occurs at a sufficiently early age, some of the functions one would expect to be lost do nevertheless develop as the child gets older. Apparently these functions have been taken over by parts of the nervous system which would not normally implement them.

Again, suppose we analyse by experiment the performance of people carrying out some common task, such as reading. We find in more and more experiments that the same apparent behaviour is being achieved by different operations by two different people in the same experiment, and even by the same person on two separate occasions. For example, reading may involve an analysis of the printed shapes directly to the meaning of the words, or a translation of each letter to a corresponding sound, with the meaning becoming available only when the sounds have been assembled. At one time these two mechanisms might have been seen as alternative theories of the way in which people read; but in fact there is good evidence that both of them occur, and that the balance between them shifts depending on the circumstances. Mental arithmetic is carried out in different ways by different people, and the types of error they make are related to the strategy they employ. There is no doubt that human functions involve the same problems we have seen in the simple computer example: the abstract description of the process may be implemented in different parts of the brain, and different sequences of operation may be used to achieve apparently similar ends. It is in fact scarcely surprising this should be so: if corporeal explanations are inadequate for a tiny machine for information processing, scarcely more than a toy, we would hardly expect them to be sufficient for a system as large and complex as a human being.

In the first half of this paper, therefore, I have been arguing against the idea that corporeal explanations are enough. In the remainder I want to turn my guns against the advocates of the incorporeal. To do this I want to take those psychological explanations which do not involve physiology, and to distinguish at least three kinds. Let me take a specific example of each kind.

First, consider the following phenomenon. If somebody plays a high-pitched tone into one of your ears, and a low-pitched into the other, you will hear two sounds coming from opposite directions, and totally different in quality. If however somebody takes the sound of an ordinary human voice, separates out all the high frequencies and puts them in one of your ears, and simultaneously takes all the low frequencies and puts them in the other ear, you will hear a single voice in the middle of your head. From a number of experiments by myself and Peter Ladefoged, conducted over twenty years ago, it is known that the key factor lies in the 'amplitude modulation' of the signals at the two ears. If you take high and low notes from a human voice, in the way I have described, the modulation frequency will be the same even though the part of the spectrum occupied is totally different. If artificially this modulation frequency is made different, then the high and low sounds split apart and you no longer hear a single voice.

We have therefore quite a satisfactory explanation relating the physical stimulus to the reported experience of the listener. There is still no satisfactory physiological account of what is going on in this situation. In the long run, I would confidently expect that we shall find something about the material nature of the auditory system, which makes it function as a detector of modulation frequencies. Indeed, I hope that our understanding of the physiology of the ear will be accelerated by knowing what the ear can do on a psychological level, rather more sophisticated than just knowing that we can tell high notes from low ones. Each science supports and enriches the other; long may they continue to do so. For many purposes however one can confine oneself to the psychological level, and say that the explanation of the experience lies in the nature of the outside stimulus which is present at that time.

Now let me take the second case. In recent studies at the Warneford Hospital, Dr Teasdale and his colleagues have found conditions under which people will recall pleasant memories from their past with a higher frequency, or more rapidly, than they will recall unpleasant memories. He has also found situations where the opposite is true. One cannot therefore appeal here solely to the momentary relationship between stimulus and experience or behaviour. The explanation lies in the previous environment in which Dr Teasdale has placed his experimental subjects. He has induced in them different states of mind, by getting them to read verbal passages appropriate to happy or unhappy moods. People who had been reading sentences such as 'Life is so full and interesting it's great to be alive!' were faster at recalling pleasant memories than were people who had been reading statements such as 'Looking back on my life I wonder if I have accomplished anything really worthwhile?' The speed for recalling unpleasant memories differed in the opposite direction, although that difference was not statistically significant.

In this case therefore the psychological explanation lay in a state of the person, which you would normally describe as a mood or feeling. The explanation of that state itself lay in past environmental conditions, which we can specify. In this case, we are of course even further from any link with events in the brain or elsewhere in the body. There may well be interesting physiological mechanisms to discover later, but on balance I think one can remain agnostic whether in the end the explanation may not be detached from particular physiological machinery rather as computer programs are. It is for instance rather ominous that only some people will show a suitable change in mood in this experiment; that may reflect something about their bodies, but equally it might not. The question can perfectly well remain wide open for a while, while we go into more detailed psychological explanation. The key point however is that at present the explanation lies in past experience, rather than the present stimulus.

Let me come now to my third type of psychological explanation. In another research going on in Oxford, we have been getting large numbers of manual workers to answer standardized questions which reveal how much they suffer from mild symptoms of psychological ill-health. On the average, we find that the level of anxiety among paced repetitive assembly workers is higher than that amongst people of the same level of pay and official status, who are doing non-repetitive and unpaced work. When one looks more closely however, one finds that it makes a difference what the worker says he is trying to get by working. Some of them say that they are working for money; what sociologists who are interested in this problem have called the 'instrumental' attitude to work. These people show no more anxiety in paced assembly than they do in situations more under their own control. The difference in anxiety between the two groups is due to those who say that they are working for 'satisfaction', which one can interpret as a desire for achievement or to give a fair return to society. In other words, any explanation of what is happening to the person needs to take into account his expressed purposes and intentions. We are even further here from any physiological connection, and can be totally agnostic whether such a connection will ever be made. What is even more to the point is that the intention of the person has no clear connection with any particular past experience, or anything of that sort. Clearly the situation cries out for further analysis of the psychology, but one would need to be pretty naïve to suppose that the differences in intention could all and invariably be traced back to some differences in past experience, such as a difference in toilet training. For the present, we have to confine ourselves to a psychological explanation. We appeal to intentions of which we have evidence at present, but which are not known to result either from the external stimulus nor from a known sequence of past stimuli.

There are therefore three kinds, at least, of psychological explanation. There are explanations from the situation, from the history, and from the internal state of the person. To clarify the differences between them, and again to remind you that there is nothing particularly mystical about this, I would like to take up parallel explanations in the case of computer programs. To do this, let us take as an example a computer program I have recently developed which shows phenomena requiring each of these three kinds of explanation. When the computer is switched on, it offers me the choice of interacting with one of three different persons: Dull, Proper, Mysterious. Whichever I choose, the machine produces an adjective which describes the behaviour of that person to me; and I can choose from a list of twelve adjectives, to describe how I wish to behave in return. The machine then gives me an adjective indicating how the person I have chosen now wishes to respond to me, I make another decision, and so on. Perhaps I should explain that I have deliberately attached no indication of sex to the three computer personalities, so that you may embody them in whatever form your imagination suggests as suitable. To give an example, let me first conduct an interaction with Dull.

Dull: Polite.
Myself: Very Polite.
Dull: Very Polite.
Myself: Very Polite.
Dull: Friendly.
Myself: Very Polite.
Dull: Very Polite.
Myself: (Perhaps I should be more forthcoming.) Friendly.
Dull: Friendly.
Myself: Very Friendly.
Dull: Affectionate.
Myself: (That's going a bit far.) Friendly.
Dull: Friendly.
Myself: (I'm getting bored with this.) Very Rude.
Dull: Very Rude.
Myself: (Well, that was just a passing mood.) Friendly.
Dull: Friendly.

It is I expect fairly obvious that Dull is reacting simply to the kind of behaviour which I show myself, with a very slight variation to greater or lesser degrees of intimacy, which is essentially random. If one wants to explain why Dull was Very Rude, the explanation is that I had been so myself. This kind of explanation is indeed non-corporeal, because as previously the same program could be run on many different machines; but the core of the explanation lies in the momentary stimulus.

Now let us consider what happens with the second possible computer person, Proper.

Proper: Polite.
Myself: Polite.
Proper: Friendly.
Myself: Very Polite.
Proper: Very Friendly.
Myself: Friendly.
Proper: Very Friendly.
Myself: (This seems to be a very promising relationship.) Affectionate.
Proper: Very Polite.
Myself: (Somewhat abashed.) Very Polite.
Proper: Very Friendly.

In this case, I am certainly getting some response to my own behaviour, but it is no longer true that the response matches what I am doing, even with a little random element mixed in. To understand what is happening, one needs to know that Proper is trying to achieve a particular value of my own behaviour: if I am Friendly, Proper will continue showing the same behaviour as immediately previously. If my attitude is less intimate than that implied by being Friendly, then Proper will show a higher degree of intimacy, with an implied belief that this will encourage me to shift towards greater warmth. If on the other hand I overstep the mark and become more than merely Friendly, then Proper will withdraw until I return to a correct degree of distance. The system is in fact showing a very simple kind of goal-seeking by negative feedback. To explain its behaviour we need to know, not merely the stimulus presented, but more importantly the state of affairs which would stop its behaviour from changing; in this case, that my behaviour should be Friendly. Without that clue, it is very difficult for an outside observer to make sense of the behaviour of Proper, because that behaviour will not be constant in response to a particular stimulus.

For example, in the preceding interaction it is obvious that I and Proper were going to settle down at a level of behaviour in which each of us was Friendly. Those with a Machiavellian turn of mind, however, may like to know that it is perfectly possible to behave in such a way that Proper will show the highest possible degree of intimacy and affection. The Machiavellian can achieve this by behaving once or twice with extreme rudeness; this will cause Proper to move immediately to the most affectionate end of the scale, and if you then simply return to the level of friendly, Proper will stay totally devoted to you.

To explain behaviour, therefore, one needs to know the goal or target; and this of course is represented inside the machine by the state of a part of its memory. The particular value of the goal has itself been determined by the program which I fed into the machine before the interaction started. While the immediate psychological explanation therefore is in the goal of the system, the more ultimate explanation lies in the program; the previous sequence of events which occurred sometime in the past.

I said a little earlier that one would need to be very naïve to suppose that everybody who works for money has one kind of toilet training, and everybody who works for satisfaction has had another. If such simplistic theories were the only ones which could be implemented in material systems such as computers or brains, then indeed one might sympathize with those who want incorporeal explanation. I suspect that some of those on that side may recognize that a computer program is an abstract sequence of events, whose physical implementation may take many different forms; but this does not resolve their misgivings, because whatever the physical machine which may be involved, they assume that the explanation of the resulting behaviour lies in the instructions with which the machine has been fed. In psychological terms, they assume that all scientific explanation must *either* be genetic, and in terms of the material structure of the system; *or* environmental in terms of its past history.

As the simplest way of showing that there are other possibilities, let me turn to my third computer person, Mysterious. In the previous two cases, since it was I who wrote the program, I was able to predict in advance how Dull and Proper would react to particular overtures of my own. In the case of Mysterious, however, it is still true that I wrote the program; but I assure you that I do not know in advance what is going to happen when the interaction begins. Historical explanation is inappropriate in this case.

Mysterious: Polite.
Myself: Polite.
Mysterious: Indifferent.
Myself: Polite.
Mysterious: Cool.
Myself: Very Cool.

Mysterious: Very Friendly.
Myself: Polite.
Mysterious: Friendly.
Myself: Very Friendly.
Mysterious: Cool.

Let me tell you what I did know before the interaction started. Mysterious works on a broadly similar system to Proper, that is, the behaviour of Mysterious is guided by a goal or desired value for my behaviour. In the case of Proper, however, this goal or target is known to anyone who knows the program. In the case of Mysterious, however, the value of the goal is only fixed during the actual running of the program, and not by the original programmer. That is why I do not know what is going to happen. In fact for ease of demonstration I have made the choice of the goal extremely simple, and Mysterious simply looks up a random number to decide what goal is to be achieved. I could have made things much more complicated, by allowing the goal to be computed in the light of a number of interactions carried out by Mysterious with other people, between my starting the program running and my own beginning to interact. Or I could have allowed the program to carry out within itself imaginary interactions with a variety of hypothetical different personalities, and to decide from those which goal would suit best some more fundamental criterion. There is no difficulty in principle about complications of this kind; but the general point is that my knowledge of the program, just like knowledge of the physical structure of the machine, is insufficient to explain the behaviour which results.

Let us now consider what kind of rational explanations would be possible in this case. If we were dealing with a human being, the clear and obvious way of getting an explanation for their behaviour would be to ask them to explain themselves. Let us try that for Mysterious.

Myself: Why? We want to know why you were Cool?
Mysterious: I wanted you to be Indifferent. You were Very Friendly, and that's too intimate, so I thought I should be more distant.

In other words, the machine itself is able in this case to tell us what its goal was. If you look back at the sequence of the behaviour, you will see that knowledge of the goal makes sense of the strange jumps and deviations which Mysterious was showing. The machine has in fact a kind of introspection, in that the crucial part of its memory, which contains its goal or purpose, has a connection with observable public events. In this way we can learn what the goal was.

To summarize, this program shows behaviour which is not predictable by the programmer, and which nevertheless can be explained by an internal state of the system. We can find out what that internal state is through a chain of events leading to a final public event. All these terms are non-corporeal, in the sense that I have been using that term; Mysterious is not confined to this particular machine and the physical events which display this behaviour could take a number of different forms. The same personality could be reincarnated in different bodies.

Now, before you run away with the idea that all psychological problems can be interpreted in terms of people's own statements about their conscious purposes, let me warn you that I have still not finished my argument. This very simple program illustrates a large number of the difficulties of explanation which occur when we are dealing with human beings. Admittedly, I have not arranged for the machine to show certain human characteristics. It will not tell deliberate lies, nor make jokes at the expense of the person who is interacting with it, but there are plenty of other ways in which trouble can arise.

First, to find out the goal which Mysterious has set, it is necessary to ask the machine what the goal is. It is not possible to assume that this system is working in the same way, and towards the same purposes, as I am myself. Many eminent authorities appear to believe that our knowledge of other minds is derived by analogy with our own. This is an extremely dangerous assumption; in practical life, one will very often encounter people who insist that yes means yes and no means no, and that you must not assume that their purposes are the same as yours. As I said earlier, the whole tenor of modern experimental research is that different minds work in different ways.

Second, communication by asking people to give an account of themselves has very serious limits. Suppose we press Mysterious further, by asking for the origins of the goal. *Why* should it be particularly desirable for my behaviour to be Indifferent? In that case we shall get only the reply 'I'm afraid I do not know the answer to that'. The limits of what can be communicated from the machine to me, or from me to you, are very tight.

Third, I am afraid we can only explain the behaviour of Mysterious retrospectively. Once we have asked what the goal is, it is changed again so that we can never use the stated intention to explain future behaviour, because the question itself alters the future. This is of course extremely common in psychological research, where people's behaviour is frequently altered by the enquiry itself.

Fourthly, if we set about asking Mysterious questions in the wrong way, we may get results which are seriously misleading, even though there is no deliberate deception going on. For example, suppose that I make you a present of the information that the goal resides in memory at a place which is identified by the letter T. (Since this is a non-corporeal statement, the physical place in memory will be different in different machines, but the symbol T will always refer to the right place.) It is therefore possible after reaching the end of the program, to obtain a printout from the machine of the value of the goal. Suppose we try it.

Myself: Print 'T'.
Mysterious: T.

We have here an example of a long-standing problem in introspective psychology, and in the philosophy of mind: the confusion between something one wishes to describe, and the term which is used to refer to it. In discussions about knowledge of other minds, I am always worried as to whether somebody is discussing the way in which we should use the term 'knowledge', or whether he is discussing relations between the concept indicated by that term and other concepts. In the case of Mysterious, where we went wrong was to use the symbol 'T', rather than T. Let's try again.

Myself: Print T.
Mysterious: 5

This reply also is difficult, because Mysterious has replied in a private internal language which is not the language we ourselves employ. The various adjectives describing social behaviour are all manipulated inside the computer in terms of numbers, and when we ask in this way for a statement of the goal, we get a reply which is true but is a number rather than a word. In the case of human psychology, the corresponding point is that you cannot take what people say at face value; they may be using words in a different way from you, or the reply to the question you ask may have quite a different significance.

In fact, you can get information about the goal of Mysterious, in the following way.

Myself: Print A $ (T).
Mysterious: Indifferent.

Here we have got a clear indication of the goal, in a language we can understand; but using a question which is not on the face of it asking what we wish to know. The analogy in human psychology is perhaps to the kind of questions we ask to reveal mental ill-health; quite often we do not take the questions at face value.

I have illustrated therefore a number of difficulties which arise when one is trying to explain the behaviour of a very small computer, running a very short program. (The program is so short that I attach it as an appendix; it is not some great achievement of computer science, but trivially simple.) As I said in the first half of this paper, explanations of the behaviour of the system have to be non-corporeal. Even with such a simple system, however, a number of difficulties arise in reaching a satisfactory explanation. The common thread in all these difficulties is that we have to establish a reliable and determinate chain of events leading from the non-corporeal entity inside the system to some visible and material event, from which we can infer what is happening. In simple cases, this event may be the stimulus, or the past history of the system. In more interesting and advanced cases, it may be some internal state which cannot be observed directly. Yet in the last case *particularly* it is important to pin down firmly what observable criteria are used to infer any particular internal state. Even in a system as simple as my computer, we cannot assume that a certain past history will produce a certain internal state, nor that a particular kind of question will get the sort of answer we want, nor that the answer will be in the terms we are used to using, nor that all the processes going on inside the system are accessible at all using our pre-established habits of language. Above all, we cannot assume that asking a question will leave the system unchanged. All these difficulties and a few more, apply in the case of human psychology also. It is certainly the case that physiological explanations of human behaviour are only part of the story, just as the particular arrangement of cards and chips in the computer has no bearing on the kinds of difficulty I have been discussing. However, to reach any certainty about the abstract events which are taking place inside people, the major source of difficulty is the drawing of a link between those events and something that we can observe. We need the corporeal connection, the evidence from experiment.

How does all this hark back to the role of the theory, the paradigm, or the presupposition which Professor Harris raised and which has echoed through the series? I have time only for one amateur suggestion; and that is to emphasise the role of the theories which are denied as well as asserted, *alternative* explanations. If we wish to explain the behaviour of Mysterious, the chance of guessing the correct theory straightaway is vanishingly small, because there are so many theoretical possibilities. We need to divide the possible theories into families, and set one family against its alternatives; to hold one theory is to deny the others. When Professor Harris talked of the river at Oxford and at Henley, the theory that the rivers were the same was being contrasted with another theory, that they were different.

In this light, the problem of Darwinism, that it is an unfalsifiable theory, becomes much easier; for Darwinism does deny preformed development of species regardless of environment, and it also denies the inheritance of acquired characteristics. It is only in a peculiar science such as physics, where most alternatives have been eliminated, that one can talk of the value of a single theory in isolation.

But that is merely a digression; my main purpose is to deny the value of incorporeal explanations, and to assert the merit of the corporeal and the non-corporeal.

Appendix

The following program, INTROS, was written by Peter FitzGerald of the Oxford Department of Experimental Psychology. It is not the one employed in the actual Herbert Spencer lecture, but shows the same external behaviour. As explained in the main text, programs written by Dr Broadbent are inferior in style to those of Mr FitzGerald, and the following will therefore be more suitable for readers who wish to make their own computer introspect.

```
100 REM  -------- INTROS --------
110 REM  This is game in which the player interacts with, and attempts
120 REM  to attune his or her responses to, a number of different
130 REM  'persons'.
140 REM     The player starts by choosing the person to interact with.
150 REM  After this the player's responses are abbreviations of the
160 REM  possible states (e.g. VP for VERY POLITE – see DATA
170 REM  list); these prompt a reply by the machine person, and so on
180 REM  until "E" is input. The player then has a choice of selecting
190 REM  a new person, ending, with summary table, or asking a ques-
200 REM  tion. The only sensible question here is "WHY?", to which
210 REM  the machine provides a meaningful answer.
```

```
220 REM     The program is fairly general, so that new persons can be
230 REM   added quite easily. Most text is output through a subroutine,
240 REM   and this gives some control over text width.
250 REM     INTROS has been implemented on a Research Machines
260 REM   380Z microprocessor, using a 9K BASIC interpreter.
270 REM   Initialise:
280 CLEAR 500: REM Clears string space.
290 RANDOMIZE: REM RND (1) will now return a random number
300 REM   between 0 and 1, from a rectangular distribution.
310 REM   Number of persons held in NP, the number of states in NS,
320 REM   the width of the text display in WDTH – these are taken
330 REM   from the first DATA line.
340 READ NP, NS, WDTH
350 DIM LAB$(NS), L2$(NS), NAME$(NP), N(NP), PLAY(NP),
          MACH (NP)
360 REM   ------------------------------------------------
370 REM   The main variables are used as follows:
380 REM   MSTATE – the integer code of the machine's response (i.e.
390 REM            that of the 'person' in use).
400 REM   NA     – the integer code of the current person.
410 REM   PSTATE – the player's response code.
420 REM   TARGET – the code of the person's target state.
430 REM   LAB$() – the names of the states (responses).
440 REM   L2$()  – the abbreviations of the states.
450 REM   NAME$() – the names of the persons.
460 REM   MACH() – the code for the current state of each person.
470 REM   N()    – the number of interactions between the player
480 REM            and each person.
490 REM   PLAY() – code of last response of player to each person.
500 REM   ------------------------------------------------
510 REM   Read into string arrays:
520 FOR J=1 TO NS: READ LAB$(J): NEXT J
530 FOR J=0 TO NS: READ L2$(J): NEXT J
540 FOR J=1 TO NP: READ NAME$(J): NEXT J
550 REM   ---> PART I – The same. <---
560 REM   Player selects person:
570 PRINT CHR$(12): REM Clears screen.
580 K$= "Choose which person to meet by entering the first"
590 K$=K$+ "letter of the name: ": GOSUB 1600
600 FOR J=1 TO NP: PRINT NAME$(J): NEXT J
```

```
610 INPUT K$
620 REM Who has been selected?
630 FOR J=1 TO NP
640   IF LEFT$(NAME$(J), 1) = K$ THEN NA=J: GOTO 670
650 NEXT J
660 PRINT "Who?": GOTO 580
670 REM Choose the person's target state:
680 PRINT: ON NA GOSUB 890,900,910
690 REM Display initial states; first, have we met NA before?
700 IF N(NA)>0 THEN GOSUB 1010 ELSE GOSUB 950
710 REM Read the player's state and analyse:
720 INPUT RESP$
730 FOR J=0 TO NS
740   IF RESP$=L2$(J) THEN CODE=J: GOTO 770
750 NEXT J
760 PRINT "Eh?": GOTO 720
770 REM Branch if "E" input:
780 IF CODE=0 THEN 1070
790 N(NA)=N(NA)+1: PSTATE=CODE
800 REM Calculate machine state (with noise):
810 ON NA GOSUB 930,940,940
820 MSTATE=INT(MSTATE+0.5)-1+INT(RND(1)*3)
830 IF MSTATE<1 THEN MSTATE=1 ELSE IF MSTATE>NS
    THEN MSTATE=NS
840 REM Output states:
850 PRINT LAB$(PSTATE): PRINT TAB(20); LAB$(MSTATE)
860 GOTO 720
870 REM --->  Routines for PART I <---
880 REM Set target:
890   RETURN: REM 'DULL' mimics.
900   TARG=8: RETURN: REM 'PROPER' wants to be friendly.
910   TARG=5+3*INT (RND(1)*3): REM 'MYSTERIOUS' is less
    predictable.
920 REM Determine the person's response:
930   MSTATE=PSTATE: RETURN
940   MSTATE=(TARG-PSTATE)+MSTATE: RETURN
950 REM Generate state for first encounter and output:
960   K$= "On meeting"+NAME$(NA)+"you find things like this:"
970   GOSUB 1600
980   PRINT 'You'; TAB (20); NAME$(NA)
```

```
990   MSTATE=5+INT(RND(1)*3): PSTATE=6: REM Latter
      arbitrary
1000  PRINT TAB(20); LAB$(MSTATE): RETURN
1010 REM Output states from last encounter:
1020  K$= "When you last met " +NAME$(NA)+ " things were like
      this:"
1030  GOSUB 1600: PSTATE=PLAY(NA): MSTATE=MACH(NA)
1040  PRINT "YOU"; TAB(20); NAME$(NA)
1050  PRINT TAB(20); LAB$(MSTATE): PRINT LAB$(PSTATE)
1060  RETURN
1070 REM ---> PART II – Questions and summary. <---
1080 PRINT: PRINT
1090 MACH(NA)=MSTATE: PLAY(NA)=PSTATE: REM Save states.
1100 K$= "Press Q for question, P for new person, or E for end."
1110 GOSUB 1600: INPUT K$
1120 IF K$="E" THEN 1140 ELSE IF K$="P" THEN 560
1130 IF K$="Q" THEN 1230 ELSE 1080
1140 REM Output summary statistics:
1150 PRINT CHR$(12): REM Clears screen.
1160 PRINT  "Name of    Trials  Person's  Your"
1170        "person             state     state"
1180        "-----      ---     -----     ---"
1190 FOR J=1 TO NP
1200  PRINT NAME$(J); TAB(14); N(J); TAB(22); MACH(J);
      TAB(32); PLAY(J)
1210 NEXT J
1220 END
1230 REM The player wants to ask a question:
1240 PRINT CHR$(12): REM Clears screen.
1250 K$= "What do you want to ask?": GOSUB 1600
1260 INPUT K$: PRINT: PRINT
1270 IF K$= "WHY?" THEN 1310 ELSE IF K$= "BUT WHY?"
      THEN 1300
1280 K$= "I'm afraid I don't know the answer to that."
1290 GOSUB 1600: GOTO 1070
1300 K$= "Why do you want that?": GOSUB 1600: PRINT: GOTO
      1280
1310 REM The machine joins with the player in expanding the question.
1320 K$= "We want to know why you were '+LAB$(MSTATE)+'.":
      GOSUB 1600
```

```
1330 REM Branch according to person:
1340 PRINT: ON NA GOSUB 1380, 1420, 1420
1350 PRINT: K$= "Do you want to ask any more?": GOSUB 1600
1360 INPUT K$: IF K$= "YES" THEN 1230 ELSE 1070
1370 REM ---> Routines for PART II. <---
1380 REM Here deal with DULL:
1390   K$= "Because I want to be like you."
1400   K$=K$+ "You were " +LAB$(PSTATE)+ " so I was much the
     same."
1410   GOSUB 1600: RETURN
1420 REM Deal here with PROPER and MYSTERIOUS:
1430   K$= "Because I wanted you to be " +LAB$(TARG)+"."
1440   K$=K$+" You were " +LAB$(PSTATE)+ "."
1450   REM . . and how did the person feel about that?:
1460   IF PSTATE>TARG THEN 1490 ELSE IF PSTATE<TARG
     THEN 1550
1470   K$=K$+" that's just right, so I didn't change too much."
1480   GOSUB 1600: RETURN
1490 REM . . . a bit forward!
1500   K$=K$+" that was too intimate so I thought I should"
1510   K$=K$+ " be more distant"
1520   REM Adjust if the person was more intimate than the player:
1530   IF MSTATE<PSTATE THEN K$=K$+".": GOSUB 1600:
     RETURN
1540   K$=K$+ " – at least more than I was before.": GOSUB 1600:
     RETURN
1550 REM . . the player was unmoved:
1560   K$=K$+ " that was too cold and I wanted to encourage you"
1570   REM Adjust if the person was cooler than the player:
1580   IF MSTATE>PSTATE THEN K$=K$+ "." GOSUB 1600:
     RETURN
1590   K$=K$+ " – at least more than I did before.": GOSUB 1600:
     RETURN
1600 REM General subroutine to give nice text output.
1610   REM K$ holds text, WDTH is display width.
1620   LK=LEN(K$): IF LK=0 THEN RETURN
1630   IF LK<=WDTH THEN PRINT K$: RETURN
1640   FOR LK=WDTH+1 TO 1 STEP -1
1650   IF MID$(K$, LK, 1) =" " THEN 1670
1660   NEXT LK
```

```
1670    IF LK>1 THEN 1700
1680    PRINT MID$(K$, 1, WDTH)
1690    K$=MID$(K$, WDTH+1): GOTO 1620
1700    PRINT MID$ (K$, 1, LK−1): K$=MID$(K$, LK+1): GOTO
    1620
1710 REM **** Now the data: ****
1720 REM Number of persons and states, and desired text width:
1730    DATA 3, 12, 39
1740 REM The name of the states (held in LAB$):
1750    DATA "VERY RUDE", "RUDE", "VERY COOL", "COOL",
    "INDIFFERENT"
1760    DATA "POLITE", "VERY POLITE", "FRIENDLY", "VERY
    FRIENDLY"
1770    DATA "AFFECTIONATE", "VERY AFFECTIONATE",
    "LOVING"
1780 REM . . and their abbreviations (held in L2$):
1790    DATA E, VR, R, VC, C, I, P, VP, F, VF, A, VA, L
1800 REM The names of the persons (held in NAME$):
1810    DATA DULL, PROPER, MYSTERIOUS
```

6 *Philosophers and human understanding*

H. PUTNAM

Harvard University

I find myself in the position that Jerome Bruner[1] found himself in a few years ago. I agreed to give a Herbert Spencer lecture; I planned to give a lecture on the topic of scientific explanation; I intended to discuss a particular controversy in that field, the controversy about whether scientific theories are 'incommensurable', about whether there is any 'convergence' in scientific knowledge; but I felt increasing dissatisfaction with this entire idea as the day approached. Bruner's dissatisfaction led him to some reflections about the history and present state of psychology. I intend to follow his example and ruminate on the activity itself, the activity of philosophy of science, in my own case, and on certain dissatisfactions I feel with the way that activity has been pursued, rather than discuss a particular issue within it. However the particular issue I mentioned will come up in the course of these ruminations.

Logical positivism is self-refuting

In the late 1920s, in approximately 1928, the Vienna Circle announced the first of what were to be a series of formulations of an empiricist meaning criterion: *the meaning of a sentence is its method of verification*. A.J. Ayer's *Language, truth and logic* spread the new message to the English-speaking philosophical world: *untestable statements are cognitively meaningless*. A statement must either be (a) analytic (logically true or logically false, to be more precise) or (b) empirically testable, or (c) *nonsense*, i.e. not a real statement at all, but only a pseudo-statement. Notice that this was already a change from the first formulation.

An obvious rejoinder was to say that the logical positivist criterion of significance was *self-refuting*: for the criterion itself is neither (a) analytic (unless, perhaps, it is analytically *false*!), nor (b) empirically testable. Strangely enough this criticism had very little impact on the logical positivists and did little to impede the growth of their movement. I want to suggest that the neglect of this particular philosophical gambit was a great mistake; that the gambit is not only correct, but contains a deep lesson, and not just a lesson about logical positivism.

The point I am going to develop will depend on the following observation: the forms of 'verification' allowed by the logical positivists are forms which have been *institutionalized* by modern society. What can be 'verified' in the positivist sense can be verified to be correct (in a non-philosophical or pre-philosophical sense of 'correct'), or to be probably correct, or to be highly successful science, as the case may be; and the public recognition of the correctness, or the probable correctness, or the 'highly successful scientific theory' status, exemplifies and reinforces images of knowledge and norms of reasonableness maintained by the culture.

The *original* positivist paradigm of verification was not this publicly institutionalized one, it is true. In Carnap's *Logische Aufbau der Welt* (*The logical construction of the world*) verification was ultimately private, based on sensations whose subjective quality or 'content' was said to be 'incommunicable'. But, under the urgings of Neurath, Carnap soon shifted to a more public, more 'intersubjective', conception of verification.

Popper has stressed the idea that scientific predictions are confronted with 'basic sentences', sentences such as 'the right pan of the balance is down' which are publicly accepted even if they cannot be 'proved' to the satisfaction of a sceptic. He has been criticized for using 'conventionalist' language here, for speaking as if it were a convention or social decision to accept a basic sentence; but I think that what some take to be the conventionalist element in Popper's thought is simply a recognition of the *institutionalized* nature of the implicit norms to which we appeal in ordinary perceptual judgments. The nature of our response to a sceptic who challenges us to 'prove' such statements as 'I am standing on the floor' testifies to the existence of social norms *requiring* agreement to such statements in the appropriate circumstances.

Wittgenstein argued that without such public norms, norms shared by a group and constituting a 'form of life', language and even thought itself would be impossible. For Wittgenstein it is absurd to ask if the institutionalized verification I have been speaking of is 'really' justificatory. In *On certainty* Wittgenstein remarks that philosophers can provide one with a hundred epistemological 'justifications' of the statement 'cats don't grow on trees'—but *none* of them starts with anything which is more sure (in just this institutionalized sense of 'sure') than the fact that cats don't grow on trees.

Sceptics have doubted not only perceptual judgments but ordinary inductions. Hume, whose distinction between what is *rational* and what is *reasonable* I am not observing, would have said there is no *rational* proof that it will snow (or even that it will probably snow) in the United States this winter (although he would have added that it would be most unreasonable to doubt that it will). Yet our response to a sceptic who challenges us to 'prove' that it will snow in the United States this winter testifies that there are social norms requiring agreement to such 'inductions' just as much as to ordinary perceptual judgments about people standing on floors and about equal arm balances (indeed, 'cats don't grow on trees' is an 'induction' in this sense.)

When we come to high-level theories in the exact sciences, people's reactions are somewhat different. Ordinary people cannot 'verify' the special theory of relativity. Indeed, ordinary people do not at the present time even *learn* the special theory, or the (relatively elementary) mathematics needed to understand it, although it is beginning to be taught in first-year physics courses in some of our colleges. Ordinary people defer to scientists for an informed (and socially accepted) appraisal of a theory of this type. And because of the instability of scientific theories, a scientist is not likely to refer to even so successful a theory as special relativity as 'true' *tout court*. But the judgment of the scientific community is that special relativity is a 'successful'—in fact, like quantum electrodynamics, an unprecedentedly successful—scientific theory, which yields 'successful predictions' and which is 'supported by a vast number of experiments'. And these judgments are, in fact, deferred to by other members of the society. The difference between this case and the cases of institutionalized norms of verification previously referred to (apart from the hedging of the adjective 'true') is the special role of experts and the institutionalized deference to experts that such a case involves; but this is no more than an instance of the division of intellectual labour (not to mention intellectual authority relations) in the society. The judgment that special relativity and quantum electrodynamics are 'the most successful physical theories we have' is one which is made by authorities which the society has appointed and whose authority is recognized by a host of practices and ceremonies, and in that sense institutionalized.

Recently it occurred to me that Wittgenstein may well have thought that *only* statements that can be verified in some such 'institutionalized' way can be true (or right, or correct, or justified) at all. I do not mean to suggest that any philosopher ever held the view that *all* things which count in our society as 'justifications' really are such. Philosophers generally distinguish between institutions which are constitutive of our concepts themselves and those which have some other status, although there is much controversy about how to make such a distinction. I mean to suggest that Wittgenstein thought that it was some subset of our institutionalized verification norms that determines what it is right to say in the various 'language games' we play and what is wrong, and that there is no objective rightness or wrongness beyond this. Although such an interpretation does fit much that Wittgenstein says—for instance, the stress on the need for 'agreement in our judgments' in order to have concepts at all—I do not feel sure that it is right. It is just too vague who the 'we' is in Wittgenstein's talk of 'our' judgments; and I do not know whether his 'forms of life' correspond to the institutionalized norms I have mentioned. But this interpretation occurred to me upon reading Wittgenstein's *Lectures and conversations*. In this Wittgenstein rejects both psychoanalysis and Darwin's theory of evolution (although unlike the positivists he does not regard such language as *meaningless*, and he has admiration for Freud's cleverness). Wittgenstein's view about psychoanalysis (which he calls a 'myth') does not signify much, although the reasons given are interesting, since so many people have the view—mistakenly in my opinion—that psychoanalysis is total nonsense. Indeed, Wittgenstein is, as I mentioned, considerably more charitable than this. But his rejection of *evolution* is quite striking.[2] Wittgenstein contrasts Darwin's theory unfavourably with theories in physics ('One of the most important things about an explanation is that it should work, that it should enable us to predict something. Physics is connected with Engineering. The bridge must not fall down.' [*Lectures on Aesthetics*, p. 25]). And he says people were persuaded 'on grounds which were extremely thin'. 'In the end you forget entirely every question of verification, you are just sure it must have been like that.'

Again, the great discussions about 'analyticity' that went on in the 1950s seem to me to be connected with the desire of philosophers to find an objective, *uncontroversial* foundation for their arguments. 'Analyticity', i.e. the doctrine of truth by virtue of meaning alone, came under attack because it had been *over-used* by philosophers. But why had philosophers been tempted to announce that so many things which are in *no* intelligible sense 'rules of language', or consequences of rules of language, were analytic or 'conceptually necessary', or whatever? The answer, I think, is that the idea that there is a definite set of *rules of language* and that *these* can settle what is and is not rational, had two advantages, as philosophers thought: (1) the 'rules of language' (or the 'depth grammar', or whatever) are constitutive institutionalized practices (or norms which underlie such practices), and as such have the 'public' status I have described; (2) at the same time, it was claimed that only philosophers (and not linguists) could discover these mysterious things. It was a nice idea while it lasted, but it was bound to be exploded, and it was.

I shall call any conception according to which there are institutionalized norms which define what is and is not rationally acceptable a *criterial* conception of rationality. The logical positivists, Wittgenstein (at least on the admittedly uncertain interpretation I have essayed) and some though not all[3] of the 'ordinary language' philosophers shared a criterial conception of rationality even if they differed on other issues, such as whether to call unverifiable statements 'meaningless', and over whether or not some ethical propositions could be 'conceptually necessary'.

The gambit I referred to at the outset, the gambit that refutes the logical positivists' verification principle, is *deep* precisely because it refutes every attempt to argue for a criterial conception of rationality, that is because it refutes the thesis that nothing is rationally verifiable unless it is criterially verifiable.

The point is that although the philosophers I mentioned often spoke as if their arguments had the same kind of *finality* as a mathematical proof or a demonstration experiment in physics; that although the logical positivists called their work *logic* of science; although the Wittgensteinians displayed unbelievable arrogance towards philosphers who could not 'see' that all philosophical activity of a pre-Wittgensteinian or non-Wittgensteinian kind is nonsensical; and although ordinary language philosophers referred to each others' arguments and those of non-ordinary language philosophers as 'howlers' (as if philosophical errors were like mistakes on an arithmetic test); no philosophical position of any importance can be verified in the conclusive and culturally recognized way I have described. In short, if it is true that only statements that can be criterially verified can be rationally acceptable, that statement itself cannot be criterially verified, and hence cannot be rationally acceptable. If there is such a thing as rationality at all—and we commit ourselves to believing in *some* notion of rationality by engaging in the activities of *speaking* and *arguing*—then it is self refuting to *argue* for the position that it is identical with or properly contained in what the institutionalized norms of the culture determine to be instances of it. For no such argument can be certified to be correct, or even probably correct, by those norms alone.

I do not at all think that rational argumentation and rational justification are impossible in philosophy, but rather I have been driven to recognize something which is probably evident to laymen if not to philosophers, namely that we cannot appeal to public norms to decide what is and is not rationally argued and justified in philosophy. The claim which is still often heard that philosophy is 'conceptual analysis', that the *concepts themselves* determine what philosophical arguments are right is, when combined with the doctrine that concepts are norms or rules underlying public linguistic practices, just a covert form of the claim that all rational justification in philosophy is criterial, and that philosophical truth is (barring 'howlers') as publicly demonstrable as scientific truth. Such a view seems to me to be simply unreasonable in the light of the whole history of the subject, including the recent history.

Let me emphasize again that I *do* think that there are rationally justified and rationally unjustified views in philosophy. I have argued at length in a whole series of publications that mind-body dualism is an unreasonable view; and I stand by that argument. My argument appealed to the great explanatory attractiveness of a 'monist' view, that is a view which does not require either property-dualism or substance-dualism, and tried at the same time to concede what was *right* in dualism (for example that belief-desire explanation is not simply 'reducible' to mechanistic explanation). In general I tried to account for dualist intuitions. But a dualist can, of course, point out counterintuitive consequences of my view; and, just as I say it is more *reasonable* to believe that we are material objects in a physical world, he can say it is more *reasonable* to believe that mental events and properties are non-identical with physical events and properties. There is no neutral place, no neutral conception of rationality, from which to decide who is right. Even if one is a neutral *on this issue*—i.e. one thinks *neither* of us has a more reasonable position than the other—*that* isn't a neutral position either. One cannot *criterially* verify that neither of us has reason on his side by appeal to the cultural norms. In fact, both of us are trying to *shape* the *future* cultural norms; so even if there *were* a culturally indoctrinated conception supporting one of us, that wouldn't convince the other.

What goes for philosophical argument goes for arguments about religion and about secular ideology as well. An argument between an intelligent liberal and an intelligent Marxist will have the same character as a philosophic dispute at the end, even if more empirical facts are relevant. And we all do have views about religion, or politics, or philosophy, and we all argue them and criticize the arguments of others. Indeed, even in 'science', outside the exact sciences, we have arguments in history, in sociology, in clinical psychology of exactly this character. It is true that the logical positivists broadened their description of the 'scientific method' to include these subjects; but so broadened it cannot be shown to clearly *exclude* anything whatsoever.

The positivists, I will be reminded, *conceded* that the verification principle was 'cognitively meaningless'.[4] They said it was a *proposal*, and as such not true or false. But they *argued* for their proposal, and the arguments were (and had to be) non-starters. So the point stands.

In sum, what the logical positivists and Wittgenstein (and perhaps the later Quine as well) did was to *produce philosophies which leave no room for a rational activity of philosophy*. This is why these views are self-refuting; and what the little gambit I have been discussing rests upon is really a very significant argument of the kind philosophers call a 'transcendental argument': arguing about the nature of rationality is an activity that *presupposes* a notion of rational justification wider than the positivist notion, indeed wider than institutionalized criterial rationality.

Anarchism is self-refuting

Let me now discuss a very different philosophical tendency. Thomas Kuhn's *The structure of scientific revolutions* enthralled vast numbers of readers, and appalled most philosophers of science because of its emphasis on what seemed to be *irrational* determinants of scientific theory acceptance and by its use of such terms as 'conversion' and 'Gestalt switch'. In fact, Kuhn made a number of important points about how scientific theories and scientific activity should be viewed. I have expressed a belief in the importance of the notions of *paradigm*, *normal science*, and *scientific revolution* elsewhere; I want to focus here on what I do *not* find sympathetic in Kuhn's book, what I described elsewhere as 'Kuhn's extreme relativism'.

The reading that enthralled Kuhn's more sophomoric readers was one on which he is saying there is no such thing as rational justification in science, its *just* Gestalt switches and conversions. Kuhn has rejected this interpretation of the *SSR*, and has since introduced a notion of 'non-paradigmatic rationality' which may be closely related to if not the same as what I just called 'non-criterial rationality'.

The tendency that most readers thought they detected in Kuhn's *SSR* certainly manifested itself in Paul Feyerabend's *Against method*. Feyerabend, like Kuhn, stressed the manner in which different cultures and historical epochs produce different paradigms of rationality. He suggests that the determinants of *our* conception of scientific rationality are largely what *we* would call irrational. In effect, although he does not put it this way, he suggests that the modern scientific-technological conception of rationality is fraudulent by its own standards. (I think I detect a similar strain in Michel Foucault.) And he goes far beyond Kuhn or Foucault in suggesting that even the vaunted instrumental superiority of our science may be somewhat of a hoax. Faith healers can do more to relieve your pain than doctors, Feyerabend claims.

It is not those terrifyingly radical claims that I want to talk about, although they are the reason Feyerabend calls his position 'anarchism'. I wish to discuss a claim Kuhn does make in both the *SSR* and subsequent papers, and that Feyerabend made both in *Against method* and in technical papers. This is the thesis of *incommensurability*. (The very thesis I originally intended to devote this lecture to.) I want to say that this thesis, like the logical positivist thesis about meaning and verification, is a self-refuting thesis. In short, I want to claim that *both* of the two most influential philosophies of science of the twentieth century, certainly the two that have interested scientists and non-philosophers generally, the only two the educated general reader is likely to have even heard of, are self-refuting. Of course, as a philosopher of science I find it a bit troublesome that this should be the case. We shall shortly come to the question of what to make of this situation.

The incommensurability thesis is the thesis that terms used in another culture, say, the term 'temperature' as used by a seventeenth-century scientist, cannot be equated in meaning or reference with any terms or expressions *we* possess. As Kuhn puts it, scientists with different paradigms inhabit 'different worlds'. 'Electron' as used around 1900 referred to objects in one 'world'; as used today it refers to objects in quite a different 'world'. This thesis is supposed to apply to observational language as well as to so-called 'theoretical language'; indeed, according to Feyerabend, ordinary language *is* simply a false theory.

The rejoinder this time is that if this thesis were really true then we could not translate other languages—or even past stages of our own language—at all. And if we cannot interpret organisms' noises at all; then we have no grounds for regarding them as *thinkers*, *speakers*, or even *persons*. In short, if Feyerabend (and Kuhn at his most incommensurable) were right, then members of other cultures, including seventeenth-century scientists, would be conceptualizable by us only as animals producing responses to stimuli (including noises that curiously resemble English or Italian). To tell us that Galileo had 'incommensurable' notions *and then go on to describe them at length* is totally incoherent.

This problem is posed in a sympathetic essay on Feyerabend's views by Smart: 'Surely it is a neutral fact that in order to see Mercury we have to point the telescope over the top of that tree, say, and not, as predicted by Newtonian theory, over the top of that chimney pot. And surely one can talk of trees, chimney pots, and telescopes in a way which is independent of the choice between Newtonian and Einsteinian theory. However Feyerabend could well concede that we use Euclidean geometry and non-relativistic optics for the theory of our telescope. He would say that this is not the real truth about our telescope, the tree, and the chimney pot, but nevertheless it is legitimate to think in this way in order to discuss the observational tests of general relativity, since we know on theoretical grounds that our predictions will be unaffected (up to the limits of observational error) if we avail ourselves of this computational convenience.'[5] But the trouble with Smart's rescue move is that I must understand *some* of the Euclidean non-relativists, language to even say the 'predictions' are the same. If *every word has a different significance*, in what sense can any prediction be 'unaffected'? How can I even translate the logical particles (the words for 'if–then', 'not', and so on) in seventeenth-century Italian, or whatever, if I cannot find a translation manual connecting seventeenth-century Italian and modern English that makes some kind of systematic sense of the seventeenth-century corpus, both in itself and in its extra-linguistic setting? Even if I am the speaker who employs both theories (as Smart envisages) how can I be justified in equating any word in my Newtonian theory with any word in my general relativistic theory?

The point I am making comes into even sharper focus when we apply to it some of Quine's and Davidson's observations about meaning and translation practice. Once it is conceded that we can find a translation scheme which 'works' in the case of a seventeenth-century text, at least in the context fixed by our interests and the use to which the translation will be put, what sense does it have *in that context* to say that the translation does not 'really' capture the sense or reference of the original? It is not, after all, as if we had or were likely to have criteria for sameness of sense or reference apart from our translation schemes and our explicit or implicit requirements for their empirical adequacy. One can understand the assertion that a translation fails to capture exactly the sense or reference of the original as an admission that a better translation scheme might be found; but it makes only an illusion of sense to say that all possible translation schemes fail to capture the 'real' sense or reference. Synonymy exists only as a relation, or better, as a family of relations, each of them somewhat vague, which we employ to equate different expressions for the purposes of interpretation. The idea that there is some such thing as 'real' synonymy apart from all workable practices of mutual interpretation, has been discarded as a myth.

Suppose someone tells us that, in certain contexts, the German word 'Rad' can be translated as 'wheel'. If he goes on to say his translation is not perfect, we naturally expect him to indicate how it might be improved, supplemented by a gloss, or whatever. But if he goes on to say that 'Rad' can be translated as 'wheel', but it does not actually refer to wheels, or indeed to any objects recognized in your conceptual system, what do we get from this? To say that a word A can be translated as 'wheel', or whatever, is to say that, to the extent that the translation can be relied upon, A *refers* to wheels.

Perhaps the reason that the incommensurability thesis intrigues people so much, apart from the appeal which all incoherent ideas seem to have, is the tendency to confuse or conflate concept and conception. To the extent that the analytic/synthetic distinction is fuzzy, this distinction too is fuzzy; but all interpretation involves such a distinction, even if it is relative to the interpretation itself. When we translate a word as, say, *temperature* we equate the reference and, to the extent that we stick to our translation, the sense of the translated expression with that of our own term 'temperature', at least as we use it in that context. (Of course, there are various devices we can use, such as special glosses, to delimit or delineate the way we are employing 'temperature', or whatever the word may be, in the context.) In this sense we equate the 'concept' in question with our own 'concept' of temperature. But so doing is compatible with the fact that the seventeenth-century scientists, or whoever, may have had a different *conception* of temperature, that is a different set of beliefs about it and its nature than we do, different 'images of knowledge', and different ultimate beliefs about many other matters as well. That conceptions differ does not prove the impossibility of ever translating anyone 'really correctly' as is sometimes supposed; on the contrary, we could not say that conceptions differ and how they differ if we could not translate.

But, it may be asked, how do we ever know that a translation scheme 'works' if conceptions always turn out to be different? The answer to this question, as given by various thinkers from Vico down to the present day, is that interpretative success does not require that the translatees' beliefs come out the *same* as our own, but it does require that they come out *intelligible* to us. This is the basis of all the various maxims of interpretative charity or 'benefit of the doubt', such as 'interpret them so they come out believers of truths and lovers of the good', or 'interpret them so that their beliefs come out reasonable in the light of what they have been taught and have experienced', or Vico's own directive to maximize the *humanity* of the person being interpreted. It is a constitutive fact about human experience in one world of different cultures interacting in history while individually undergoing slower or more rapid change that we are, as a matter of universal human experience, able to *do* this; able to interpret one another's beliefs, desires, and utterances so that it all makes some kind of *sense*.

Kuhn and Feyerabend, not surprisingly, reject any idea of *convergence* in scientific knowledge. Since we are not talking about the same things as previous scientists, we are not getting more and more knowledge about the same microscopic or macroscopic objects. Kuhn argues that science 'progresses' only instrumentally; we get better and better able to transport people from one place to another, and so on. But this too is incoherent. Unless such locutions as 'transport people from one place to another' retain some degree of fixity of reference, how can we understand the notion of instrumental success in any stable way?

The argument I have just employed is essentially related to Kant's celebrated argument about preconditions for empirical knowledge. Replying to the contention that the future might be wholly lawless, might defeat every 'induction' we have made, Kant pointed out that if there is any future at all—any future *for us*, at any rate—any future we can grasp as thinkers and conceptualize to see if our predictions were true or false—then, in fact, many regularities must *not* have been violated. Else why call it a *future*? For example, when we imagine balls coming from an urn in some 'irregular' order, we forget that we *could not even tell they were balls*, or tell *what order they came out in*, without depending on many regularities. *Comparison* presupposes there are some commensurabilities.

There is a move Kuhn and Feyerabend could make in reply to all these criticisms, although it is not one they would feel happy making, and that would be to introduce some kind of observational/theoretical dichotomy. They could concede commensurability, translatability, and even convergence with respect to observational facts, and restrict the incommensurability thesis to the theoretical vocabulary. Even then there would be problems (why should we not describe the meanings of the theoretical terms *via* their relations to the observational vocabulary *à la* Ramsey?) But Kuhn and Feyerabend reject this alternative with reason, for in fact the need for principles of interpretative charity is just as pervasive in 'observational language' as in 'theoretical language'. Consider, for example, the common word 'grass'. Different speakers, depending on where and when they live have different perceptual prototypes of grass (grass has different colours and different shapes in different places) and different conceptions of grass. Even if all speakers must know that grass is a plant, on pain of being said to have a different concept altogether, the conception of a *plant* today involves photosynthesis and the conception of a plant two hundred years ago did not. Without interpretative charity which directs us to equate 'plant' 200 years ago with 'plant' today (at least in ordinary contexts) and 'grass' 200 years ago with 'grass' today, no statement about the reference of this word 200 years ago could be made. Nor is it only natural kind words that are so dependent for interpretation on principles of charity; the artefact word 'bread' would pose exactly the same problems. Indeed, without interpretative charity we could not equate a simple colour term such as 'red' across different speakers. We interpret discourse always as a whole; and the interpretation of 'observation' terms is as dependent on the interpretation of 'theoretical' terms as is the interpretation of the latter on the former.

What I have given is, once again, a transcendental argument. We are committed by our fundamental conceptions to treating not just our present time-slices, but also our past selves, our ancestors, and members of other cultures past and present, as *persons*; and that means, I have argued, attributing to them shared references and shared concepts, however different the *conceptions* that we also attribute. Not only do we share objects and concepts with others, to the extent that the interpretative exercise succeeds, but also conceptions of the reasonable, of the natural, and so on. For the whole justification of an interpretative scheme, remember, is that it renders the behaviour of others at least minimally reasonable by *our* lights. However different our images of knowledge and conceptions of rationality, we share a huge fund of assumptions and beliefs about what is reasonable with even the most bizarre culture we can succeed in interpreting at all.

What to make of this?

The two arguments I have just set out convinced me that the two most widely known philosophies of science produced in this century are both incoherent. (Of course, neither of them is *just* a 'philosophy of science'.) This naturally led me to reflect on the meaning of this situation. How did such views arise?

Logical positivism, I recalled, was both continuous with and different from the Machian positivism which preceded it. Mach's positivism, or 'empirio-criticism', was, in fact, largely a restatement of Humean empiricism in a different jargon. Mach's brilliance, his dogmatic and enthusiastic style, and his scientific eminence made his Positivism a large cultural issue (Lenin, afraid that the Bolsheviks would be converted to 'empirio-criticism', wrote a polemic against it). Einstein, whose interpretation of special relativity was operationalist in spirit (in marked contrast to the interpretation he gave to general relativity), acknowledged that his criticism of the notion of simultaneity owed much to Hume and to Mach, although, to his disappointment, Mach totally rejected special relativity.

But the most striking event that led up to the appearance of logical positivism was the revolution in deductive logic. By 1879 Frege had discovered an algorithm, a mechanical proof procedure that embraces what is today standard 'second order logic'. The procedure is *complete* for the elementary theory of deduction ('first order logic'). The fact that one can write down an algorithm for proving *all* of the valid formulas of first order logic—an algorithm which requires no significant analysis and simulation of full human psychology—is a remarkable fact. It inspired the hope that one might do the same for so-called 'inductive logic'—that the 'scientific method' might turn out to be an algorithm, and that these two algorithms—the algorithm for deductive logic (which, of course, turned out to be *incomplete* when extended to higher logic) and the algorithm-to-be-discovered for inductive logic—might exhautively describe or 'rationally reconstruct' not just *scientific* rationality, but all rationality worthy of the name.

When I was just starting my teaching career at Princeton University I got to know Rudolf Carnap, who was spending two years at the Institute for Advanced Studies. One memorable afternoon, Carnap described to me how he had come to be a philosopher. Carnap explained to me that he had been a graduate student in physics, studying logic in Frege's seminar. The text was *Principia mathematica* (imagine studying Russell and Whitehead's *Principia* with Frege!) Carnap was fascinated with symbolic logic and equally fascinated with the special theory of relativity. So he decided to make his thesis a formalization of special relativity in the notation of *Principia*. It was because the Physics Department at Jena would not accept this that Carnap became a philosopher, he told me.

Today, a host of negative results, including some powerful considerations due to Nelson Goodman, have indicated that there *cannot* be a completely *formal* inductive logic. Some important aspects of inductive logic can be formalized (although the adequacy of the formalization is controversial), but there is always a need for judgments of 'reasonableness', whether these are built in via the choice of vocabulary (or, more precisely, the *division* of the vocabulary into 'projectible' predicates and 'non-projectible' predicates) or however. Today, virtually no one believes that there is a formalizable scientific method, one that can be completely formalized without formalizing complete human psychology (and possibly not even then).

The story Carnap told me supports the idea that it was the success of formalization in the special case of deductive logic that played a crucial role. If that success inspired the rise of logical positivism, could it not have been the failure to formalize inductive logic, the discovery that there is no *algorithm* for empirical science, that inspired the rise of 'anarchism'? I am not a historian, so I will not press this suggestion. In any case, additional factors are at work. While Kuhn has increasingly moderated his view, both Feyerabend and Michel Foucault have tended to push it to extremes. Moreover there is something political in their minds: both Feyerabend and Foucault link our present institutionalized criteria of rationality with capitalism, exploitation, and even with sexual repression. Clearly there are many divergent reasons why people are attracted to extreme relativism today, the idea that all existing institutions and traditions are bad being one of them.

Another reason is a certain *scientism*. The scientistic character of logical positivism is quite overt and unashamed; but I think there is also a scientism hidden behind relativism. The theory that all there is to 'rationality' is what your local culture says there is is never quite embraced by any of the 'anarchist' thinkers, but it is the natural limit of their tendency: and this is a reductionist theory. That rationality is defined by an ideal computer program is a scientistic theory inspired by the exact sciences; that it is simply defined by the local cultural norms is a scientistic theory inspired by anthropology.

I will not discuss here the expectation aroused in some by Chomskian linguistics that cognitive psychology will discover *innate* algorithms which define rationality. I myself think that this is an intellectual fashion which will be disappointed just as the logical positivist hope for a symbolic inductive logic was disappointed.

All this suggests that part of the problem with present-day philosophy is scientism inherited from the nineteenth century—a problem that affects more than one intellectual field. I do not deny that logic is important, or that formal studies in confirmation theory, in semantics of natural language, and so on are important. I do tend to think that they are rather peripheral to philosophy, and that as long as we are too much in the grip of formalization we can expect this kind of swinging back and forth between the two sorts of scientism I described. Both sorts of scientism are attempts to evade the issue of giving a sane and humane description of the scope of human reason. Let me soften the rather portentous tone of this last remark by suggesting a good title for a philosophy book: 'An essay concerning human understanding'. Seriously, human understanding *is* the problem, and philosophers *should* try to produce essays and not scientific theories.

Non-criterial rationality

If we agree that rationality (in the wide sense, including Hume's 'reasonableness') is neither a matter of following a computer program nor something defined by the norms of the culture, or some subset of them, then what account *can* we give of it?

The problem is not without analogues in other areas. Some years ago I studied the behaviour of natural kind words, for example, *gold*, and I came to the conclusion that here too the extension of the term is not simply determined by a 'battery of semantic rules', or other institutionalized norms. The norms may determine that certain objects are *paradigmatic examples* of gold; but they do not determine the full extension of the term, nor is it impossible that even a paradigmatic example should turn out not to really be gold, as it would be if the norms simply *defined* what it is to be gold.

We are prepared to count something as belonging to a kind, even if our *present* tests do not suffice to show it is a member of the kind, if it ever turns out that it has the same essential nature as (or, more vaguely, is 'sufficiently similar' to) the paradigmatic examples (or the great majority of them). What the essential nature is, or what counts as sufficient similarity, depends both on the natural kind and on the context (iced tea may be 'water' in one context but not in another); but for gold what counts is ultimate composition, since this has been thought since the ancient Greeks to determine the lawful behaviour of the substance. Doubtless Locke had something like this in mind when he said we could see the 'real essences' of things if we were only able to 'see with a microscopical eye'. And, unless we say that what the ancient Greeks meant by *chrysos* was *whatever had the same essential nature* as the paradigmatic examples, then neither their search for new methods of detecting counterfeit gold (which led Archimedes to the density test) nor their physical speculations will make sense.

It is tempting to take the same line with rationality itself, and to say that what determines whether a belief is rational is not the norms of rationality of this or that culture, but an *ideal theory* of rationality, a theory which would give necessary and sufficient conditions for a belief to be rational in the relevant circumstances in any possible world. (Such a theory would tell us what property rationality 'rigidly designates', in Kripke's sense.) Such a theory would have to *account for* the paradigmatic examples, as an ideal theory of gold accounts for the paradigmatic examples of gold; but it could go beyond them, and provide criteria which would enable us to understand cases we cannot presently see to the bottom of, as our present theory of gold enables us to understand cases the most brilliant ancient Greek could not have understood.

One suggestion has already been advanced as to the content such a theory might have. Basing himself on the model of causal theories of knowledge, Alvin Goldman has suggested[6] that, in essence, what makes a method of arriving at beliefs *justificatory* is just that the method tends to produce *true* beliefs. But this suggestion, I think, cannot be right.

Part of the reason I think Goldman's idea cannot be right is that truth itself, on my view, is an idealization of rational acceptability. The idea that we have some notion of truth which is totally independent of our idea of rational acceptability just seems untenable to me. But I do not wish to defend or even presuppose the claim that it is untenable today.

Independently of considerations about the correct conception of truth, it seems quite clear that a belief can be rational (or 'justified') even though the method by which it was arrived at will not in fact lead to true beliefs in the future and even though it turns out it did not really lead to true beliefs in the past. Goldman himself concedes that when we discuss counterfactual cases (other 'possible worlds') we sometimes say people had justified beliefs even though the methods by which they arrived at them are not reliable *in those worlds*. It seems to me that Goldman has confused a genetic explanation of the origin of the notion with an elucidation of the concept of justification. It may be that we would not have the concept of justification that we do have if certain methods were not reliable; but this does not mean that *given* the concept we have, there is a necessary connection between being justified and being arrived at by a method which is reliable. It is not impossible to have a fully justified belief even though the method by which that belief was arrived at is in fact highly unreliable; there might not be any *reason* to think the method was anything but reliable. In addition, explicating 'method' will, I suspect, pose exactly the same problems as explicating 'projectibility'; and this is the rock on which inductive logic foundered.

Goldman's suggestion aside, a general difficulty with the proposal to treat 'rational', 'reasonable', 'justified', etc. as natural kind terms is that the prospects for actually *finding* powerful generalizations about all rationally acceptable beliefs seem so poor. There are powerful universal laws obeyed by all instances of gold, which is what makes it possible to describe gold as the stuff that will turn out to obey these laws when we know them; but what are the chances that we can find powerful universal generalizations obeyed by all instances of rationally justified belief? The very same considerations that defeated the program of inductive logic, the need for a criterion of 'projectability' or a 'prior probability metric' which is 'reasonable' by a standard of reasonableness which seems both topic-dependent and interest-relative, suggests that (1) the theory of rationality is not separable from our ultimate theories about the nature of the various things which make up both ourselves and the domains being investigated; and (2) even in a restricted domain, for example physics, nothing like precise laws which will decide what is and is not a reasonable inference or a justified belief are to be hoped for.

This does not at all mean that there are *no* analogies between scientific inquiry into the nature of gold and moral inquiry or philosophical inquiry. In ethics we start with judgments that individual acts are right or wrong, and we gradually formulate maxims (*not* exceptionless generalizations) based on those judgments, often accompanied by reasons or illustrative examples, as for instance 'Be kind to the stranger among you, because you know what it was like to be a stranger in Egypt'. These maxims in turn affect and alter our judgments about individual cases, so that new maxims supplementing or modifying the earlier ones may appear. After thousands of years of this dialectic between maxims and judgments about individual cases, a philosopher may come along and propose a moral conception, which may alter both maxims and singular judgments and so on.

The very same procedure may be found in all of philosophy (which is almost coextensive with theory of rationality). In a publication a few years ago I described the desiderata for a moral system, following Grice and Baker, and I included (1) the desire that one's basic assumptions, at least, should have *wide appeal*; (2) the desire that one's system should be able to withstand rational criticism; (3) the desire that the morality recommended should be *livable*. It is striking that these same desiderata, unchanged, could be listed as the desiderata for a methodology or a system of rational procedure in *any* major area of human concern.

But an analogy is not an identity. We should and must proceed in a way analogous to the way we proceed in science (thinking of 'judgments about individual cases' as analogous to 'basic sentences', 'maxims' as analogous to 'low-level generalizations', and 'conceptions' as analogous to 'theories'); but we cannot reasonably expect that *all* determined researchers are destined to converge to one moral theory or one conception or rationality.

But, as David Wiggins has reminded us recently,[7] this is not such a bad position to be in. For, indeed, how much is there ultimately that 'all determined researchers' are destined to converge upon? We cannot expect total convergence on political conceptions, on interpretations of history or even of specific historical events, on sociological conceptions or descriptions of specific institutions, any more than we can on moral conceptions or philosophical conceptions generally. Even such notions as 'earthquake' or 'person' may be discarded by or not easily translatable into the language of some cultures.

To one philosophical temperament this is sure proof that there is 'no fact of the matter' about any of these subjects; about what is right or wrong, about philosophical conceptions themselves, about liberal versus Marxist interpretation of social phenomena, or even, according to Quine, about any intentional phenomena at all—no fact of the matter, that is, about any statement involving *belief*, *desire*, or *meaning*. But I have already pointed out the self-refuting character of such a view.

To me, it seems rather that the problem, if it is a problem, has no solution—and that *is* the solution. The correct moral to draw is not that nothing is right or wrong, rational or irrational, true or false, and so on, but, as I said before, that there is no neutral place to stand, no external vantage point from which to judge what is right or wrong, rational or irrational, true or false. But is this not relativism after all? Or am I saying that each of us is a prisoner of his private conception of rationality, that for each of us there is no difference between 'justified' and 'justified by *my* lights') (a species of solipsism).

Let me speak to the question of solipsism first. I have already rejected the thesis of incommensurability. But not only are we mutually intelligible beings; we are as mutually dependent cognitively as we are materially. Whatever our differences, we all depend on others for data and, since they cannot be separated from the data, interpretations. Examples of the vast extent of this dependence could easily be given in every area. Not only do we depend on persons with whom we have a variety of disagreements for data and interpretations, but we also rely on others for corroboration. The claim that something is *true* typically involves the one who makes it in the further claim that that something could be corroborated by other rational persons; this is not empty in practice, because there are generally independent tests for the requisite sort of competence. Psychologists have pointed out that even in the simplest cases of perceptual judgment, we will change our minds if others fail to corroborate the judgment: and this is as it should be. Far from showing a distressing spinelessness in our culture, it shows a healthy respect for the good sense of others.

Even where I will stand fast against others—for example, in thinking that torture is totally wrong—the *application* of the principle in question is dependent upon data and interpretations from others. I would not let others convince me that torture is sometimes right, but I would listen to and be influenced by the opinions and perceptions of others in deciding whether there was torture in a given case, and what the degree of moral guilt was.

As for relativism, that comes in two forms. There is the old relativist view that since there are a plurality of cultures and conceptions, therefore they are all equally good. This is really a self-contradictory view, since it is inconsistent to have a conception of rationality and simultaneously deny that that conception is better than any other. And there is the newer relativism that seeks to show that *our* conception of rationality is fraudulent by its own lights, a mere rationalization of transient and repressive institutions. I have already said that I do not agree with this. We can differ on fundamental questions, even on questions of methodology, and still listen to arguments, consider each other's assumptions and inferences, and so on. It is true that some judgments of 'reasonableness' must be made simply on the basis of ultimate intuition; not everything can be proved. And it is true that here responsible and careful thinkers may disagree, and may even consider that the other accepts illegitimate arguments. But this is very far from saying that our attempts to be rational are a fraud.

Let me close with a picture. My picture of our situation is not the famous Neurath picture of science as the enterprise of reconstructing a boat while the boat floats on the open ocean, but it is a modification of it. I would change Neurath's picture in two ways. First, I would put ethics, philosophy, in fact the whole culture in the boat, and not just 'science', for I believe all the parts of the culture are inter-dependent. And, second, my image is not of a single boat but of a *fleet* of boats. The people in each boat are trying to reconstruct their own boat without modifying it so much at any one time that the boat sinks, as in the Neurath image. In addition, people are passing supplies and tools from one boat to another and shouting advice and encouragement (or discouragement) to each other. Finally, people sometimes decide they do not like the boat they are in and move to a different boat altogether. (And sometimes a boat sinks or is abandoned.) It is all a bit chaotic; but since it is a fleet, no one is ever totally out of signalling distance from all the other boats. We are not trapped in individual solipsistic hells (or need not be) but invited to engage in a truly human dialogue; one which combines collectivity with individual responsibility.

Notes

1. Psychology and the image of man. In *Scientific models and man: The Herbert Spencer Lectures 1976* (ed. H. Harris). Oxford University Press (1979).

2. Concerning evolution what Wittgenstein said was 'People were certain on grounds which were extremely thin. Couldn't there have been

an attitude which said: "I don't know. It is an interesting hypothesis which may eventually be well confirmed".' Lectures on aesthetics. In *L. W. Wittgenstein*: *Lectures and Conversations* (ed. Cyril Burret) p. 26. University of California Press (1967). What it would be like for evolution to be 'well confirmed' Wittgenstein does not say, but the paragraph suggests that actually *seeing* speciation occur is what he has in mind ('Did anyone see this process happening? No. Has anyone seen it happening now? No. The evidence of breeding is just a drop in the bucket.')

It is instructive to contrast Wittgenstein's attitude with Monod's:

the selective theory of evolution, as Darwin himself had stated it, required the discovery of Mendelian genetics, which of course was made. This is an example, and a most important one, of what is meant by the content of a theory, the content of an idea . . . [A] good theory or a good idea will be much wider and much richer than even the inventor of the idea may know at his time. The theory may be judged precisely on this type of development, when more and more falls into its lap, even though it was not predictable that so much would come of it. (J. Monod, On the molecular theory of evolution. In *Problems of scientific revolution: progress and obstacles to progress in the sciences* (ed R. Harré). Oxford University Press (1975).)

3. One might develop an 'ordinary language' philosophy which was not committed to the public and 'criterial' verification of philosophical theses if one could develop and support a conception in which the norms which govern linguistic practices are not themselves discoverable by ordinary empirical investigation. In 'Must we mean what we say', Stanley Cavell took a significant step in this direction, arguing that such norms can be known by a species of 'self knowledge' which he compared to the insight achieved through therapy and also the the transcendental knowledge sought by phenomenology. While I agree with Cavell that my knowledge as a native speaker that certain uses are deviant or non-deviant is not 'external' inductive knowledge—I can know without evidence that in my dialect of English one says 'mice' and not 'mouses' – I am inclined to think this fact of speaker's privileged access does not extend to *generalizations* about correctness and incorrectness. If I say (as Cavell does) that it is part of the rule for the correct use of locutions of the form *X is voluntary* that there should be something 'fishy' about X, then I am advancing a *theory* to explain my intuitions about specific cases, not just reporting those intuitions. It is true that something of this sort also goes on in pyschotherapy; but I am not inclined to grant self-knowledge any kind of immunity from criticism by others, including criticisms which depend on offering rival *explanations*, in either case. And if one allows the legitimacy of such criticism, then the activity of discovering such norms begins to look like social science or history—areas in which, I have argued, traditional accounts of 'The Scientific Method' shed little light. (See my *Meaning and the moral sciences*. Routledge, London (1978).)

In any case, whatever their status, I see no reason to believe that the norms for the use of *language* are what decide the extension of 'rationally acceptable', 'justified', 'well confirmed', and the like.

4. The weakest argument offered in defence of the verification principle construed as a proposal was that it 'explicated' the 'pre-analytic' notion of meaningfulness. (For a discussion of this claim, see my 'How not to talk about meaning', in my *Mind, language and reality*, *Philosophical papers*, *Vol.* 2. Cambridge University Press (1975).) Reichenback defended a form of the verification principle (in *Experience and prediction*) as *preserving all differences in meaning relevant to behaviour*. Against an obvious objection (that the non-empirical belief in a divinity —Reichenbach used the example of Egyptian cat-worshippers—could alter behaviour, Reichenback replied by proposing to translate 'Cats are divine animals' as 'cats inspire feelings of awe in cat-worshippers'. Clearly the acceptance of this substitute would *not* leave behaviour unchanged in the case of a cat worshipper!

The most interesting view was that of Carnap. According to Carnap, *all* rational reconstructions are proposals. The only factual questions concern the logical and empirical consequences of accepting this or that rational reconstruction. (Carnap compared the 'choice' of a rational reconstruction to the choice of an engine for an aeroplane.) The conclusion he drew was that in philosophy one should be tolerant of divergent rational reconstructions. However, this Principle of Tolerance, as Carnap called it, *presupposes* the verification principle. For the doctrine that no rational reconstruction is uniquely *correct* or corresponds to the way things 'really are', the doctrine that all 'external questions' are without cognitive sense, *is* just the verification principle. To apply the Principle of Tolerance to the verification principle itself would be circular.

5. J. J. C. Smart, Conflicting views about explanation. In *Boston studies in the philosophy of science Volume II: in honor of Philipp Frank* (ed. R. Cohen and M. Wartofsky). Humanities Press, New York (1965).

6. A. Goldman, What is justified belief. In *Justification and knowledge* (ed. George S. Pappas) (forthcoming).

7. See his Truth, invention, and the meaning of life. *Proceedings of the British Academy*, Vol. LXII, esp. pp. 362–3 (1976).

Index

Adler, F. 20
analytic propositions 3, 12
anarchism 105-6
Aristotle 18
Ayer, A. J. 99

Bargmann, S. 26
Bartlett, F. C. v
belief attribution 53-4, 58-62
Berlin, Sir I. 14, 18
Besso, M. 14
Beveridge, Sir W. v
Beyerchen, A. D. 52
biology 39, 42-4, 47
Bohr, N. 4, 14, 15, 22, 24, 27, 39
Born, M. 6, 22, 29
Brans, C. B. 30-1
Broadbent, D. v, ix, 76-98
Broglie, L. V. de 5
Bruner, J. 99
Büchner, L. 20

Carnap, R. 100, 112, 120
Cavell, S. 119
Cherniak, C. 74
Churchland, P. 74
circulation of the blood 42-3
circulation of lymphocytes 43-4
Columbus, R. 42, 52
Compton, A. H. 5
conventionalism 38
Copernicus 18, 29
corporeal explanation 76-8, 93

Darwin, C. 119
Davidson, D. 74, 107
Davisson, C. J. 6
Democritus 74
Dennett, D. C. vi, ix 53-75
DeVoto, B. 5, 7
Dicke, R. H. 30-1
dualism 3-6, 12, 104

Eddington, A. S. 6
Ehrenhaft, F. 7, 10, 11
Einstein, A. v, 1-4, 6-7, 11-26, 28-34, 39, 40
electromagnetism 19, 23, 31-4
Eötvös, R. 30
ether 6, 28

falsification 6, 7, 36-40, 49
Faraday, M. 25, 32
Fermi, E. 5
Feyerabend, P. vi, 36, 37, 39, 51, 73, 106-10
Fichte, J. G. 19
Fitzgerald, P. 79-81, 93
Fodor, J. 79
Foucault, M. 106, 112
Frank, P. 1, 20, 25
Franklin, B. 7
Frege, G. 111, 112
Freud, S. 21

Galileo 2
Galton, F. v
Germer, L. H. 6
Ginsberg, M. v
Goldman, A. 114-15, 120
Goodman, N. 112
Gowans, J. L. 43-4, 52
gravitation 6, 19, 23, 29-31, 34, 39
Grice, H. P. 116
Grossmann, M. 19

Haas, de 4
Haldane, J. B. S. v
Hampshire, S. 66
Harris, H. v, ix, 36-52, 92-3
Harvey, W. 39, 42-4, 45-6, 48, 52
Hayes, P. 75
Heisenberg, W. C. 4, 14, 22, 23
Hertz, H. R. 25, 32
Hilbert, D. 21
Holton, G. v, x, 1-27, 52
Hooke, R. 77
Hopf, L. 20
Hume, D. 2, 36, 44, 51, 101, 111, 113

incommensurability 25, 99, 106
incorporeal explanation 77-8, 83, 93
induction 14, 36, 44, 101, 111-12
intentional strategy 55-73
interpretationism 54, 63

Kant, I. 2, 109
Kaufmann, W. 6, 11
Kemble, E. C. 4
Kemmer, N. 33
Kepler, J. 14, 16, 27, 29-31
Klein, F. 21
Kripke, S. 114
Kuhn, T. S. vi, 36, 39, 51, 52, 78, 105-10, 112

Ladefoged, P. 83
Lakatos, I. 38, 39, 52
Laplace, P. 56
laws, 11, 21, 29, 39, 45-6
Leibnitz, G. W. 29
Levy Bruhl, L. v
Lindemann, F. 25
Locke, J. 114
logical positivism 99-100, 103-5, 106, 111-13
Lorentz, H. A. 24, 27

McCarthy, J. 75
Mach, E. 2, 17, 20, 21, 111
Mark, H. F. 11
Maugham, W. Somerset 53
Maxwell, J. C. 24, 25, 27, 31, 32
Medawar, P. B. v
Mendel, G. 39, 48
Mendeleev, D. I. 17
Mertz, J. T. 19
Miller, G. 79
Millikan, R. A. vi, 7-11, 17, 26, 27
Minkowski, H. 21
Monod, J. 39, 119

Needham, J. 23
Neurath, O. 100, 118
Newton, I. 19, 27, 29-31, 39, 45
Newtonian physics 12, 107
non-corporeal explanation 76-98
Northrop, F. S. C. 3
Nozick, R. 64

Oersted, H. C. 19
Oppenheimer, R. 5

parsimony 13, 15, 23, 29
Pasteur, L. 39, 48
Petzold, J. 21
phenomenic propositions 3, 12
physics 1-35, 39, 101, 115
Planck, M. 5, 6, 11, 14, 18, 20, 21, 27, 29, 39
Poincaré, H. 2
Popper, K. v, 7, 36, 37, 39, 44, 51, 52, 78, 100
positivism 3, 79, 105, 111
Putnam, H. vi, x, 70, 74, 99-120

quantum mechanics 5, 15, 22, 101
Quine, W. V. 37, 47, 48, 51, 52, 73, 105, 107, 117

rationality 46, 101, 106, 113-18
realism 54, 67
Reichenbach, H. 120
relativism 117-18; *see also* incommensurability
relativity
  general theory 6-7, 12, 22, 24
  special theory 19, 21, 101
Riolan, J. 43
Russell, B. v, 18, 112
Rutherford, E. 27

Salam, A. v, x, 28-35
Schelling, F. W. J. 19
Schilpp, P. 26
Schleiermacher, F. D. E. 19
Schrödinger, E. 4, 5
Seelig, C. 17
Shapiro, I. 30
Shoemaker, S. 74
Smart, J. J. C. 107, 120
Spencer, H. v, 71

Spengler, O. 21
Stack, M. 74
Stich, S. 74
Stodola, A. 20
suspension of disbelief 5, 7–11, 17, 23

Teasdale, J. D. 84

Ullian, J. S. 48, 52
unification (of science) 2, 13, 15, 17–23, 28, 32

verification 36, 40–4, 49, 99–105
Vico, G. 109
Von Humboldt, F. H. A. 19
Von Laue, M. 20

Weinberg, S. 16
Weltbild 19–24
Whitehead, A. N. 112
Wiggins, D. 116
Whitteridge, G. 52
Wittgenstein, L. 100–3, 105, 118, 119